The
Welder's Bible

Second edition

The Welder's Bible

Second edition

Don Geary

TAB Books
Division of McGraw-Hill, Inc.
New York   San Francisco   Washington, D.C.   Auckland   Bogotá
Caracas   Lisbon   London   Madrid   Mexico City   Milan
Montreal   New Delhi   San Juan   Singapore
Sydney   Tokyo   Toronto

© 1993 by **TAB Books**.
TAB Books is a division of McGraw-Hill, Inc.

Printed in the United States of America. All rights reserved. The publisher takes no responsibility for the use of any of the materials or methods described in this book, nor for the products thereof.

pbk   8 9 10 11 FGR/FGR 9 9 8
hc 1 2 3 4 5 6 7 8 9   FGR/FGR 9 9 8 7 6 5 4 3 2

Library of Congress Cataloging-in-Publication Data

Geary, Don.
    The welder's bible / by Don Geary.—2nd ed.
       p.   cm.
    Includes index.
    ISBN 0-8306-3826-1   ISBN 0-8306-3825-3 (pbk.)
    1. Oxyacetylene welding and cutting—Amateurs' manuals.
I. Title.
TS228.G42 1992
671.5′2—dc20   92-1677
                                     CIP

Acquisitions Editor: Stacey Pomeroy
Editor: Peter D. Sandler
Director of Production: Katherine G. Brown
Book Design: Jaclyn J. Boone
Cover Design: Al Kane, Marion, Pa.
Cover Photography: Brent Blair Photography, Baltimore, Md.

TAB1
4116

Contents

Introduction

*I*n this book you will learn the basics of the oxyacetylene process for joining and cutting metals. I should mention, at this point, that I do not advise the do-it-yourselfer to teach himself or herself how to weld with any method other than oxyacetylene. While the text is designed for the do-it-yourself welder, it contains many descriptions of professional welding and cutting techniques. I trust you will find these additions helpful in learning how to weld. If any terms are unfamiliar to you, check the glossary. I have tried to make it as complete and helpful as possible.

Working with oxyacetylene welding or cutting equipment can be dangerous. Always approach any welding problem soberly; use your best judgment. Safety should always be your first priority. In fact, there is an entire chapter devoted to safety. Read the chapter on safety before setting up your welding equipment or lighting your torch for the first time.

Though it is entirely possible to become a proficient welder solely with the aid of this book and many hours of practice, you should supplement the information contained in these pages with advice and instruction from a professional welder. Spend the time to seek out professionals in your area. These people are often warehouses of both practical and technical information. If time permits, you should also enroll in an educational program that deals specifically with welding and cutting. The cost is usually minimal and you get the opportunity to learn things that cannot possibly be found in a welding book.

If you do not own an oxyacetylene welding or cutting torch, but are considering buying one, buy the best you can afford. I suggest that you visit a welding supply house in your area. Look in the yellow pages. Often, the people who work at supplying welders with tools and materials are very helpful to beginners. Currently, a basic welding and cutting outfit costs about $200. Used equipment, in good condition, might cost half as much. To determine if used welding equipment is in good condition, have a professional who specializes in welding equipment check it over.

It takes many hours to learn how to join or cut metal with the oxyacetylene process. Take the time. Learn how to do things properly.

Chapter **1**

Oxyacetylene welding fuels

Welding is a relatively new way of joining two or more pieces of metal in order to make the finished piece as strong as the original metal. The oldest type of welding, *oxyacetylene welding*, was developed around the turn of the twentieth century. There have been many developments in metal joining processes since then.

For our purposes, metal joining can be broken down into three rather broad categories: *gas welding, electric welding,* and *gas/electric welding.* Because the home welder is not interested in production work, there is little need, other than for informational proposes, to learn about other types of welding such as *laser* or *plasma welding.* These processes are used solely in industry.

HISTORY OF GAS WELDING

Gas welding is the oldest of all types of welding and the easiest to learn. It is very simple in principle. Basically, oxygen and acetylene are burned together to produce a flame that is hotter than the melting point of most metals. The temperature of an oxyacetylene flame is generally accepted as being around 6,000 degrees Fahrenheit.

Because oxyacetylene is so widely used, it is almost unbelievable that this process did not come into existence until the end of the nineteenth century. Oxyacetylene welding was first made possible through the experiments and discoveries of a French chemist, Le Chatelier, in 1895. Le Chatelier was the first to discover that burning oxygen and acetylene produced a flame with a temperature far higher than any other flame.

It wasn't long before the capabilities of oxyacetylene became known to the industrial world and were put to use. After a workable way to store and transport oxygen and acetylene was developed, the road was clear for widespread use of this new method of joining metals.

World War I probably accelerated the use of oxyacetylene welding. The pressures to supply a fighting army and to repair heavy equipment brought oxyacetylene welding in contact with millions of people. After the war, there was a need for greater controls over the welding process, and machines were developed that could weld.

The oxyacetylene welding/cutting process is the most versatile means of working with metals. No other equipment or process used by the metal industries is capable of performing such a wide variety of work on most types and thicknesses of metals. The oxyacetylene method of welding is also the easiest to master and probably the most versatile for the do-it-yourself welder. The bulk of this book concentrates on this type of welding.

Actually the oxyacetylene process can be used for joining, heating, and cutting metals. Joining or fusion welding is an important application of the oxyacetylene process: the edges of two pieces of metal are heated up to their melting points and fused together (FIG. 1-1).

Heating with the oxyacetylene process is often done for forming metals into various shapes and for other heat-treating operations such as annealing, flame hardening, tempering, case hardening, and stress relieving (FIG. 1-2).

1-1 Welding is a process in which separate pieces of metal are heated up to the melting point and fused together.

I-2 Welding is often used to heat up and shape metals, such as this automobile fender.

The last important use of oxyacetylene is in cutting metals. A stream of pure oxygen is directed against an area of heated metal. This action causes the metal to oxidize or burn and thus be cut (FIG. 1-3).

As the name implies, oxyacetylene welding uses a combination of oxygen and acetylene. To understand the welding process, it is helpful to know these two substances.

OXYGEN

About one-fifth of the air we breathe is pure oxygen. By contrast, oxygen used in the welding process is about as pure as possible—over 99 percent pure. The method of producing pure oxygen for welding and medical purposes is called the *liquid-air process*.

Liquid-air process

To oversimplify the liquid-air process, about 20 percent of our atmospheric air, as mentioned earlier, is pure oxygen, 78 percent is nitrogen, and 2 percent consists of other gases. Oxygen and nitrogen have different boiling temperatures. Thus, it is easy to separate the two by heating atmospheric air to a certain temperature and holding it at this temperature until

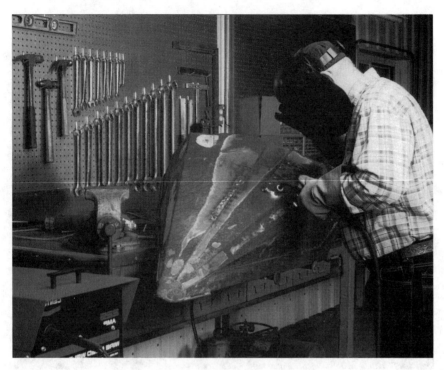

I-3 Welding equipment is also used to cut metals.

the nitrogen, which has a boiling point of 295 degrees Fahrenheit, boils off. After the nitrogen has been removed from atmospheric air, oxygen and a small mixture of other gases remain. These include carbon dioxide, argon, hydrogen, neon, and helium. Because oxygen has the highest boiling point of all these gases, the remaining mixture is further heated until only pure oxygen remains. The pure oxygen is then stored as either a gas or a liquid, depending on the eventual use. The liquid-air process is probably the most widely used way to produce pure oxygen.

Cylinders

Oxygen is commonly sold in cylinders in three sizes: 244 cubic feet, 122 cubic feet, and 80 cubic feet (FIG. 1-4). There are very strict requirements for oxygen cylinders. Each must be able to withstand more than a ton of pressure per square inch. The Interstate Commerce Commission (ICC) has set up guidelines for oxygen cylinders. No part of the cylinder may be less than 1/4 inch thick. Each cylinder must be made or forged from a single piece of steel. The steel itself must be armor plate, high-carbon steel (FIG. 1-5).

Because the ICC requires periodic inspection of oxygen cylinders, which are used as shipping containers, very few oxygen cylinders are actually owned by private individuals. The oxygen supply houses, which

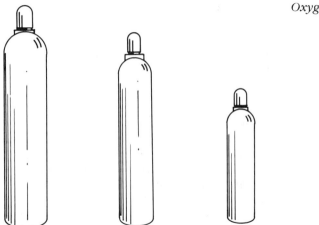

I-4   The three most popular sizes of portable oxygen cylinders, left to right, are 244 cubic feet, 122 cubic feet, and 80 cubic feet.

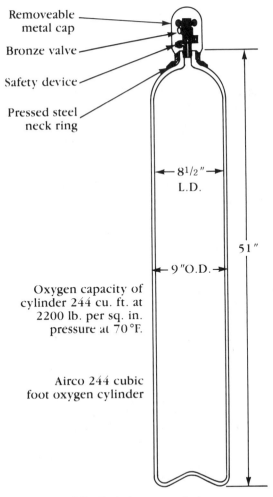

Removeable metal cap

Bronze valve

Safety device

Pressed steel neck ring

8¹/₂″ L.D.

9″ O.D.

51″

Oxygen capacity of cylinder 244 cu. ft. at 2200 lb. per sq. in. pressure at 70°F.

Airco 244 cubic foot oxygen cylinder

I-5   Typical oxygen cylinder.

own and lease oxygen cylinders, have the responsibility of complying with the ICC regulations and inspections. This makes the whole affair easier on the consumer.

I have found that you can lease oxygen, as well as acetylene, cylinders. When you require more oxygen, you simply return the empty cylinder to the dealer, and he replaces it with a full one. All you pay for is the oxygen. The oxygen itself is what you use. In most parts of the country, a 20-year lease is available at a reasonable rate. In my case, the cost of a cylinder of oxygen and a cylinder for acetylene totals about $10 a year. Oxygen and acetylene, are bought as needed and are extra.

The universally accepted color for oxygen cylinders, lines, and control knobs is green. Because there is no regulation requiring oxygen cylinders to be green in color, many companies paint their oxygen cylinders a special identifying color. It will be to your advantage to become familiar with the color used by your dealer.

Safety

Oxygen cylinders are not dangerous when used and stored according to stated safety precautions. Nevertheless, some general comments are in order.

Use only cylinders carrying ICC markings. You can be certain that cylinders of this type comply with the stringent regulations of the ICC.

Store cylinders only in a safe location, and fasten them in place. This practice ensures that the cylinder will not be knocked over. Keep tanks away from stoves, radiators, furnaces, or other overly warm places. Oxygen cylinders should also be kept away from all combustible materials or liquids. If cylinders are stored in the open, they should be protected from the elements of water, heat, cold, and the sun's direct rays.

Never use oxygen cylinders for any purpose other than holding oxygen. You should never use oxygen cylinders as rollers to help move large or heavy objects, for example. Nor should you use oxygen cylinders as supports.

Oxygen cylinders should be stored and used in an upright position. In most cases a hand truck, specifically designed to hold two cylinders (both oxygen and acetylene), is the best means for storing and using oxygen. The cylinders should be securely chained or strapped to the cart in order to prevent the cylinders from ever falling over and to make sure they are safely transported (FIG. 1-6).

Never use the valve on top of a cylinder to lift a cylinder from a horizontal to a vertical position. The best way to lift a cylinder is to first make sure that the valve protection cap is secured tightly, and then to raise the cylinder by grasping the cap firmly and lifting (FIG. 1-7).

Never allow oxygen cylinders to come in contact with live electrical wires or other electrical equipment. Keep cylinders away from welding and cutting work. Make certain that the hoses containing oxygen and acetylene do not lie under the work.

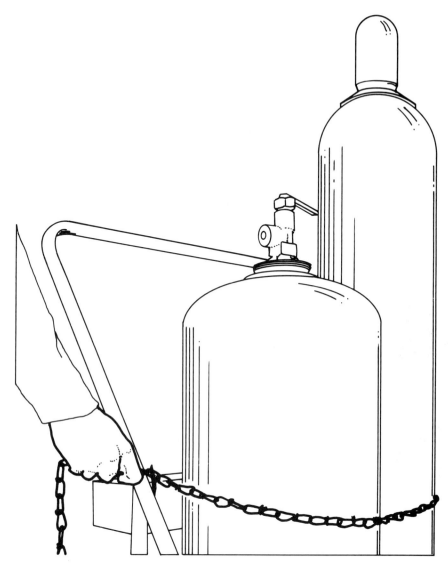

1-6 Both oxygen and acetylene tanks should be chained to prevent them from accidentally falling.

Open cylinder valves all the way when working, and always close cylinder valves when you have finished working. Never leave the cylinder valve open when you are not in the immediate vicinity. Cylinder valves should always be tightly closed when not in use—whether they are empty or full. If you stop working for lunch, for example, turn off the valves and bleed the lines. I will explain this procedure later (FIG. 1-8).

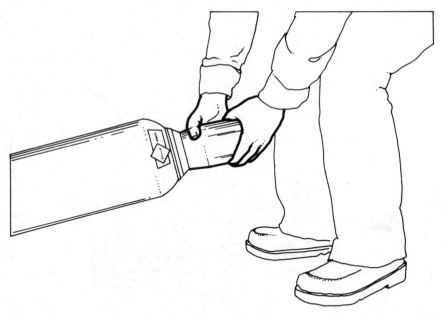

1-7   Whenever you pick up a cylinder of oxygen or acetylene, always lift it by the protective cap.

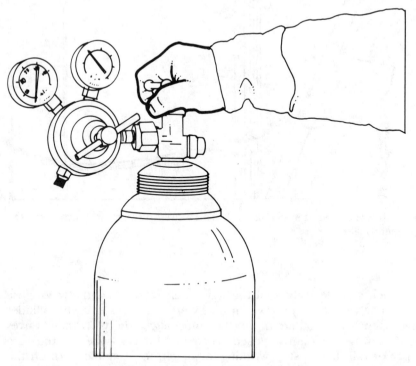

1-8   Always turn off the flow of both oxygen and acetylene at the tank when you stop work for more than 10 minutes.

Never use oxygen around oil or grease, as these burn violently in the presence of pure oxygen under pressure. Do not use any oil or grease on the regulator fitting.

Never use oxygen for anything other than welding and cutting. Oxygen is flammable. There will be a very real danger if you use it to "dust off" work or clothing. Similarly, oxygen should never be used for ventilation, pressure tests of any kind, or any other purpose.

ACETYLENE

Acetylene is produced by combining calcium carbide and water. When calcium carbide, commonly called *carbide* in the trade, is dropped into water, a reaction occurs that causes gas bubbles to rise. The gas is acetylene. It has a peculiar order and, if lighted, burns with a black, smoky flame. After the action of the carbide has ceased, a whitish residue remains in the water. This end product is hydrated or slaked lime (calcium hydroxide) and has many uses in the fertilizer industry.

As a side note, old-time miner headlamps are called *carbide* lamps and work if you simply add water to carbide pellets. A gas is formed in a special container and is forced out of a jet. This gas is ignited in front of a reflector plate, which throws a bright beam of light (FIG. 1-9).

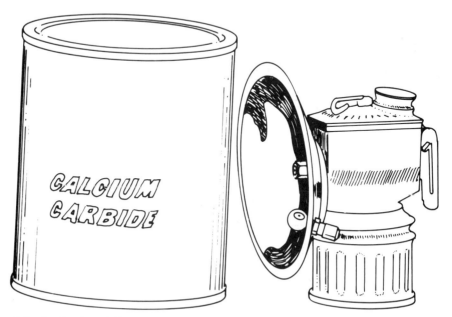

1-9 A miner's carbide lamp works on the same basic principle as an acetylene generator.

Carbide-to-water generators

Commercial production of acetylene is basically the same as a miner's headlamp, only on a much larger scale. Companies that use great amounts of acetylene will usually have their own generator, which supplies them with all the acetylene they can use. These generators are called "carbide-to-water generators" and are almost totally automated. In this type of generator, small amounts of calcium carbide are fed into a large, sealed container of water. The heat given off as the reaction occurs is absorbed by the water. The acetylene gas is captured when it rises to the top of the tank. This gas is then either used directly or stored for future use (FIG. 1-10).

Containers

For the home welder, the only practical way to obtain acetylene is in a special tank similar to, but much smaller than, a standard oxygen cylinder.

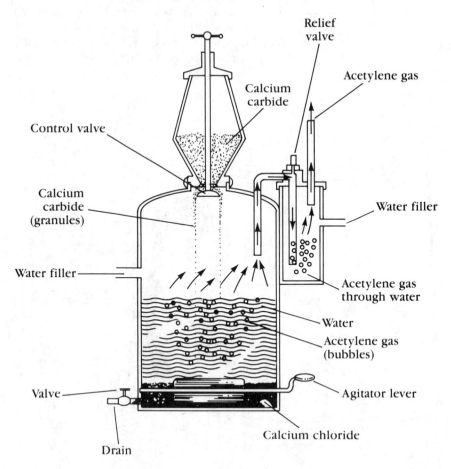

1-10   Cutaway view of an acetylene generator.

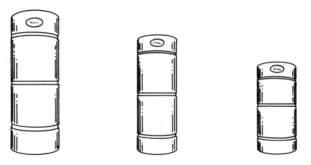

I-II The three most popular sizes of portable acetylene cylinders.

The three common sizes of acetylene cylinders are 300, 100, and 60 cubic feet. A full tank of acetylene will register approximately 225 pounds per square inch (psi) on a pressure gauge (FIG. 1-11).

As with oxygen, the ICC has set up guidelines for acetylene containers. It is against ICC regulations to store free acetylene at pressures greater than 15 psi. All welding and cutting with the oxyacetylene process can and must be done at pressures of less than 15 psi.

In order to store acetylene below the 15-psi requirement, special tanks have been developed. It is interesting to note how acetylene is stored. After a tank has been made of heavy-gauge steel, it is packed with a porous substance such as pith from cornstalks, fuller's earth, lime silica, or other materials. Next, the tank is filled with *acetone*, a liquid chemical having the ability to absorb about 25 times its own volume of acetylene. When the cylinder is filled to about 225 psi with acetylene, the acetone dissolves the acetylene and makes it safe because the acetylene is not free (FIG. 1-12).

Acetylene cylinders have a valve on top, to which the regulator is attached. This valve is similar to the valve on oxygen cylinders except that it has left-handed threads. You must turn the valve counterclockwise to open and clockwise to close. This turning action is exactly opposite to that of a standard control valve.

Remember that all acetylene connections on the cylinder, hoses, and torch handle are left-handed. Each brass fitting has a groove cut around its circumference for quick identification (FIG. 1 13).

All acetylene tanks have safety plugs on the top or bottom of the cylinder. These plugs are designed to melt, in the event of a fire, around 212 degrees Fahrenheit. Needless to say, the cylinders should be stored and used away from any heat source. Once these safety plugs melt, all the acetylene and acetone in the cylinder are allowed to escape (FIG. 1-14).

Precautions

As with oxygen cylinders, there are a few general safety precautions that should always be followed for safe operation and storage of acetylene.

Always call acetylene by its proper name—*acetylene*. It should never

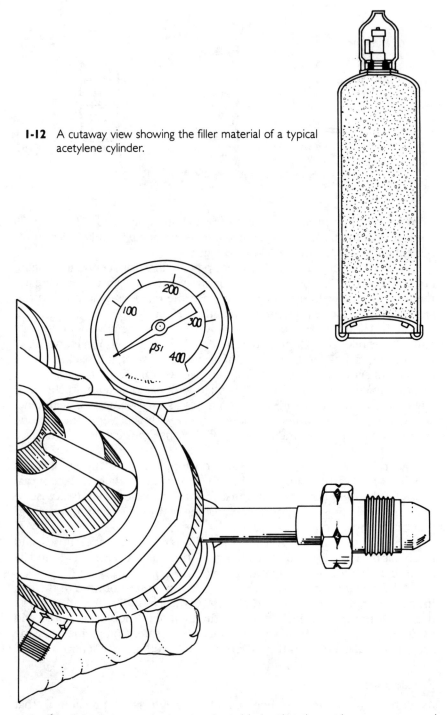

1-12   A cutaway view showing the filler material of a typical acetylene cylinder.

1-13   The fitting is for an acetylene regulator. Notice that the nut has grooves around the middle. All left-handed threaded fittings have this groove.

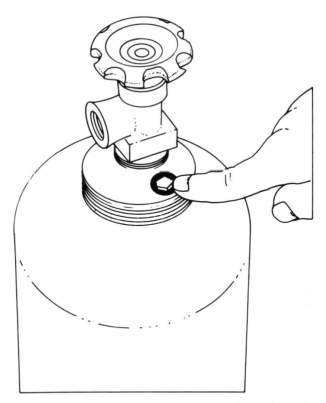

1-14 The safety plug on an acetylene cylinder is specially designed to melt at temperatures above 212 degrees Fahrenheit.

be called *gas*, as it is very different from the type of gas used in a kitchen, furnace, or automobile.

Acetylene should be stored away from all heat sources and other fuels. Ideal storage conditions include a cool, dry area away from all combustible materials.

Handle acetylene cylinders with special care. The safety plugs melt at about the same temperature as boiling water and can "blow out" as a result of rough handling. As with oxygen, cylinders of acetylene should be secured in such a way that the cylinder cannot fall over. A specially designed hand truck is probably the best way to move and store acetylene and oxygen.

Do not open the valve on an acetylene cylinder more than one and one-half turns as this can cause some of the acetone in the tank to escape with the acetylene. Acetone will damage all rubber and plastic parts in the system including hose and regulator. Therefore, it should not be allowed to escape.

Always use acetylene in an upright position. Never use it when the cylinder is in any other position than vertical. Acetone might escape along with the acetylene.

Chapter **2**

Oxyacetylene welding equipment

*T*his chapter examines the important parts of the oxyacetylene welding system. Always use equipment properly. Proper use of equipment results in safe and better heating, welding, and cutting.

REGULATORS

Oxygen and acetylene are commonly used from special cylinders and a full tank of oxygen tips the scales at about 2,200 psi, and a tank of acetylene will measure about 225 psi. Most welding and cutting can be done with a maximum of 50 psi for oxygen and 10 psi for acetylene. Obviously, some type of device must be attached to the tank to reduce the high pressure of the tank to a lower working pressure. In addition, this device must provide a steady flow of oxygen or acetylene over the life of the tank. The pressure in a full tank is obviously more than in a half-full tank, but the working pressure will be the same. Another requirement of such a device is that it must regulate consistently the flow of oxygen or acetylene from the tank to the torch.

There are basically two different types of regulators for the oxyacetylene welder: single-stage regulators and two-stage regulators.

Single-stage regulator

A single-stage regulator, as the name implies, reduces the pressure coming out of a cylinder to a desired working pressure in one step. Single-stage regulators are less expensive than two-stage regulators. They are practical for general welding, cutting, and heating. A single-stage regulator shows more of a pressure rise or drop as the cylinder contents are used than will a two-stage regulator. Generally, this deviation in pressure does not affect the quality of the flame on small, short-term work. However, you will need to readjust the torch controls when welding for a prolonged period (FIG. 2-1).

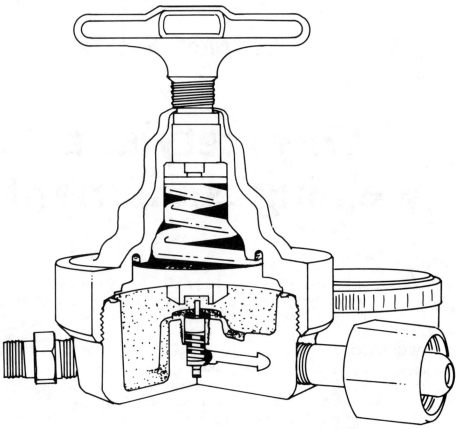

2-1 Oxygen regulator.

Two-stage regulator

A two-stage regulator maintains practically constant flow pressures during a long period of operation. Two-stage regulators are more expensive than single-stage regulators, but offer much more precise flow regulation. Generally, all two-stage regulators have two gauges; one shows the pressure in the tank and the other shows the working pressure in the line.

Actually, the term *two-stage* means that the pressure in a tank is reduced in two steps before it is allowed into the line. The first stage serves as a high pressure reduction chamber, indicating the pressure in the tank. The second stage is, in effect, a low-pressure reduction chamber and is adjusted by means of a special control knob on the face of the regulator.

Many types of regulators are produced in this country. All of them, fall into either a single- or two-stage classification. Many of these regulators have two gauges. One gauge indicates the pressure in the cylinder, and the other indicates the working pressure being delivered to the torch (FIG. 2-2). Other regulators will have only one gauge, usually showing the

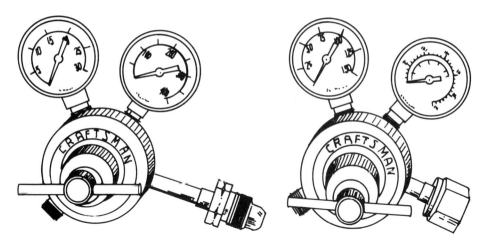

2-2 These two-stage regulators have gauges that show working pressure in the lines (left gauge) and pressure in the tank. The regulator on the left side is for acetylene and the other is for oxygen.

working pressure being supplied to the torch. These single-gauge regulators are most commonly single-stage regulators. There are also regulators that do not have any gauges. Generally, these gaugeless regulators are used in commercial work and are attached to a pipeline supplying oxygen and acetylene to a workstation (FIG. 2-3).

Although all gauges are constructed of sturdy materials, they can be damaged by carelessness or improper use. Internal working parts commonly consist of gears and springs that operate the dial indicator on the face of the unit. Exercise care when handling, installing, or removing any pressure-regulator gauge. In addition, store gauges in a safe, dust-free environment for long life.

Differences

Probably the greatest difference between a single- and two-stage regulator is that a single-stage regulator requires the torch to be adjusted as the pressure in the tank is lessened. One other objection to single-stage regulators is that they have a tendency to freeze in cold weather. A sudden expansion and resulting drop in initial pressure causes the gas to cool very rapidly. The end result is a frozen single-stage regulator.

There is a major difference between a regulator used for oxygen and one used for acetylene. Acetylene brass fittings are left-handed (and marked) and, therefore, cannot be used on oxygen tanks. Oxygen regulators are considerably stronger overall because of the pressure of standard oxygen tanks. A full tank contains approximately 2,200 psi of pressure, while a full acetylene tank contains only 225 psi.

Because regulators are probably the most important part of an oxy-acetylene welding unit, they should be taken care of properly. An acetylene regulator is designed to handle and control the high pressure of

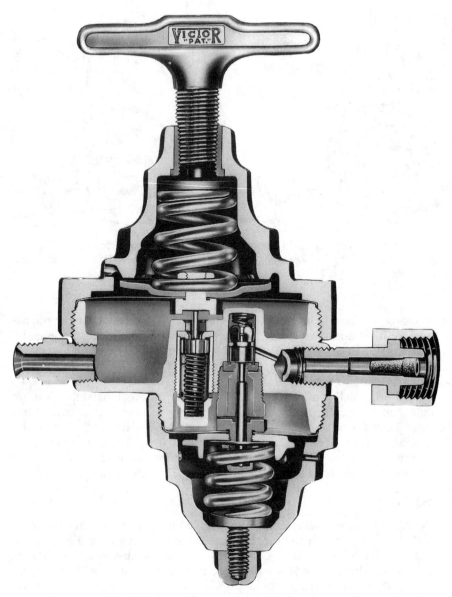

2-3 Typical acetylene regulator without gauge.

acetylene and make using this gas safe. Oxygen regulators perform a similar function at much higher pressures. Both types of regulators provide years of dependable service if they are cared for according to accepted procedures. Although specific instructions are included with all new welding regulators, there are some general comments that should be mentioned.

Care of regulators

Never use any oil on an oxygen regulator, as oil is easily ignited when it comes in contact with pure oxygen. This means that you should never oil any of the regulator parts (FIG. 2-4). Almost all regulators will have bold printing on their gauges stating "USE NO OIL."

Before attaching a regulator to an oxygen or acetylene cylinder, crack the valve. Cracking means opening the valve slightly for a second. This clears the cylinder outlet of any dirt that might have accumulated since the tank was last used. Removing any dirt or dust ensures that these particles do not get inside the regulator where they can cause the regulator to malfunction (FIG. 2-5).

After the tank valve has been cracked, the regulator can be attached. Hand-tighten the regulator, being careful not to cross-thread the brass fitting. After the fitting is snug, tighten it with a wrench. With the regulator turned off, slowly open the valve on the cylinder. It is important that you do not open the cylinder valve too quickly. The pressure from the tank

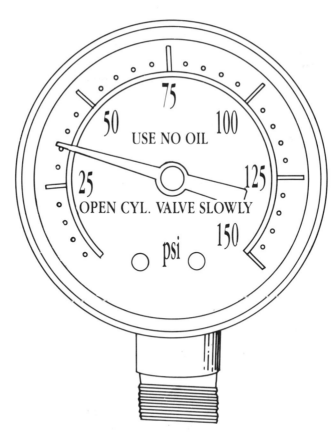

2-4 Never use any lubricant on regulator parts. Most regulators say "USE NO OIL" on the dial face.

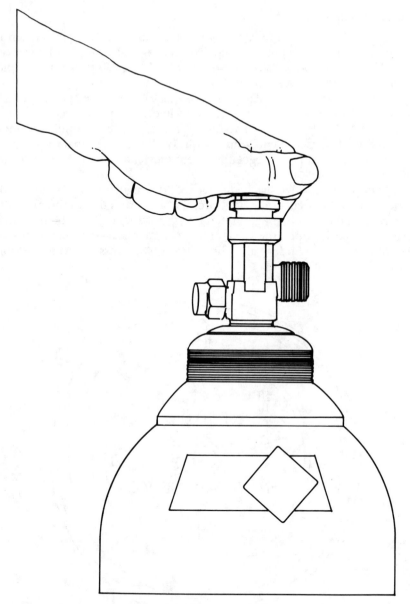

2-5 Before attaching the oxygen regulator to the tank, open the valve for a moment to clear the orifice. This is known as ''cracking'' the valve.

can damage the regulator if it is allowed into the regulator all at once. After opening the valve, check for leaks around the fitting.

Probably the best way to determine that connections are snug is to use the soap test. Simply mix up a small amount of dish soap and water. Put a few drops on the exposed threads of the fitting. If there is a leak, the soap mixture will bubble. If the connection is tight, there will be no bubbles (FIG. 2-6).

Leaks around regulators are often the result of not tightening the fitting enough. In some cases, there might be a small particle of dirt on the regulator nipple, preventing a tight seal. Remove the regulator and wipe the fitting with a clean, dry cloth. Always turn off the tank valve before tightening a loose or leaking connection.

When regulators will not be used for a period of several weeks, you should relieve the pressure on the pressure valve seat. After the regulator has been disconnected from the cylinder, turn the pressure adjusting screw until it moves freely. In most cases this will mean screwing the adjusting handle out from the body of the regulator. The reason for doing this is to relieve the pressure on the valve seat and prolong the life of the regulator unit (FIG. 2-7).

The regulator valve seat should be kept clean at all times. Some regulator units come with plastic covers, which are fitted over the regulator intake and outlet orifices. These do a lot to keep the regulator connecting fittings clean. In the absence of these covers, wipe the threads of the fittings clean with a dry cloth before attaching to the tank and fastening hose connections. While wiping these fittings, it is also a good idea to inspect them for signs of wear or damage to the threads. Damage to the threads prevents a secure connection, and the unit will leak. If this hap-

2-6 To check for leaks, the soap bubble test is the most reliable. Simply put some of the foam on the connection and watch for large air bubbles—which would indicate a leak. If no large bubbles appear, the connection is sufficiently tight.

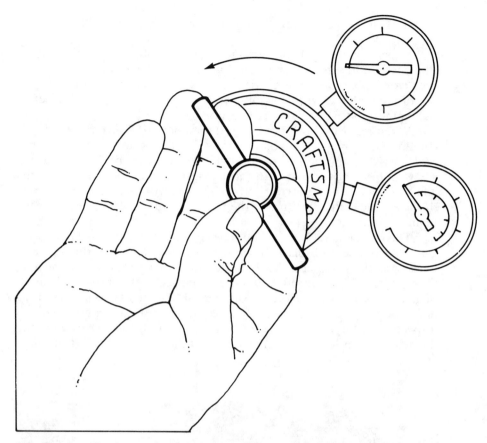

2-7　To relieve the pressure on the regulator diaphragm, turn the control lever to the left until there is little resistance.

pens, the unit should be returned to the manufacturer or qualified welding equipment repair shop. Never use a regulator in need of repair.

There are few problems with a well-maintained regulator. It will provide years of dependable service if it is simply kept clean and not dropped or otherwise damaged. Nevertheless, there are a few things that might develop through general use.

Regulator freezing

Because a regulator controls high-pressure gas, the unit can become so cold that a layer of frost collects on the surface of the regulator. As the gas expands, it absorbs heat from the regulator and cools it. As the regulator becomes cooler, the moisture in the atmosphere condenses on the cold metal parts of the regulator and freezes.

Although frost on the outside of the regulator will not affect its operation, the gas flow might fluctuate, and the flame might go out. If the regulator freezes, it usually is a direct result of moisture in the gas and not a malfunction of the regulator.

Jump

Jump is a fairly common regulator occurrence, but it is not really a problem at all. Jump usually happens when the flow of oxygen or acetylene is turned off at the torch handle. At this time, the gauge indicating the amount of pressure in the tank rises momentarily and then stops. This short pressure rise is the jump. The amount of jump really indicates the true difference between pressure at the regulator outlet when oxygen or acetylene is flowing and when it is not flowing. When the torch valve is closed, pressure on the regulator's gauge rises until the diaphragm is moved far enough for the regulator valve to close tightly. This condition of a closed valve regulator is called *lockup*. Jump is not a regulator malfunction. It is a normal reaction to changes in flow or no-flow pressures.

Creep

Creep is probably the only true problem with regulators. Creep can happen when the torch valve is shut off at the torch handle. The low-pressure or working-pressure gauge then moves upward. Limited creep is very similar to jump and might be hard to distinguish. In extreme cases, creep causes the gauge to rise until either it registers the same pressure as the cylinder or a connection is blown, usually in the hose. Creep always occurs when the regulator is defective and cannot reach lockup. Needless to say, you should never operate a regulator that has a tendency to creep.

REGULATOR SAFETY RULES

The following list can be used as a quick review on the important safety considerations covered in this chapter:

- Crack open the valve on the cylinder before installing any regulator.
- Inspect regulator and valve threads for damage before connecting.
- Be certain that you are installing the proper gauge on the cylinder at hand. As a rule, acetylene connections have a groove machined on the outside while oxygen nuts are plain. Additionally, oxygen gauges commonly have green markings, and acetylene markings are red.
- Snug up all fittings with the proper-size wrench—never use pliers as these could damage the brass nuts.
- Open the pressure-adjusting screws all the way out before opening the cylinder valve. This will ensure that damage does not happen to the internal parts of the gauges.
- Never use oil or grease around oxygen equipment. This includes hands, gloves, and clothing as well.
- Open cylinder valves slowly to prevent a surge of pressure to regulator parts. It is wise to turn valves from the side rather than the top. If the valve or gauge is faulty, it will blow up from the cylinder

top. Rapid opening results in tremendous pressure and heat on regulator parts and body.

- Check for leaks with the soapy water test, or by closing the tank
  valve and watching for a pressure drop at the gauge. If there are
  leaks the gauge reading will fall.

- Never use a regulator in need of repair. Broken glass or a tendency
  to creep is a good reason for taking your regulator to a qualified
  repair shop.

- Do not exceed the recommended working pressures for both oxygen and acetylene. Generally speaking, oxygen pressure should
  not exceed 70 psi, acetylene 15 psi.

- Install reverse flow check valves in the regulator outlet or torch
  inlet for both oxygen and acetylene connections.

HOSES

Hoses are another important part of the oxyacetylene welding system.
There are always two hoses, one for oxygen and another for acetylene.
Each end of the hose has a fitting that attaches to either the regulator or
the torch handle. As with all other connections, the oxygen fittings are
right-handed threads and the acetylene connections are left-handed with
an identifying groove cut around the circumference of the brass fitting.
All hoses are colored—green for oxygen and red for acteylene. In addition, it is common practice with new equipment for the hose maker to
print "oxygen" on the green oxygen hose. It is impossible to attach the
acetylene hose to the oxygen fitting on either the regulator or torch handle because the threads are different (FIG. 2-8).

Most hoses produced today are made from a high-quality synthetic or
natural rubber material. Oxygen and acetylene hoses are flame- and oil-
resistant and reinforced with a nylon mesh inside the hose. The end result
is a very tough hose that is tested at the factory to withstand pressures of
up to 400 psi (FIG. 2-9).

Most welding hoses in use today are double, or *Siamese*, as they are
called in the trade. This simply means that the two hoses are joined along
their entire length with a special adhesive. Handling a Siamese hose is
much easier than if two separate hoses were being used. Hoses for welding are available in several different lengths and inside diameters, the most
common being 20 feet long with a 3/16-inch inside diameter.

A quality welding hose can last for many years with a minimum of
care. Before using a new hose, the talcum powder, which all new hoses
are lined with, should be blown out with compressed air. Failure to do so
might cause some of this powder to enter the regulator or torch handle
where it can cause a malfunction. Probably the easiest way to do this is to
bring the new hose to your local gas station and use the air hose to blow
out the lines, unless, of course, you have access to compressed air in your
workshop.

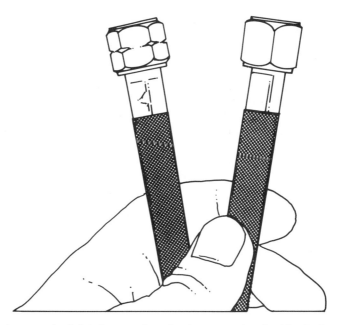

2-8 The hose on the left is for acetylene (notice grooved nut) while the hose on the right is for oxygen.

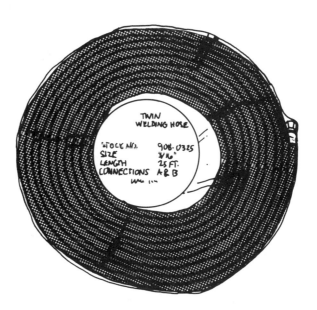

2-9 Hoses for welding are specially designed to withstand high pressure.

Periodically check your welding hose for signs of wear or deterioration. It is a good idea to wipe the hoses regularly with a clean, dry cloth. In cases of extremely dirty hoses, you can wipe them with a damp cloth. As you are cleaning the hoses, check for signs of wear.

Older hoses can be checked for leaks by immersing them in a pail of water while they are under working pressure. The same soap and water test mentioned earlier can be used to check connections around fittings and other parts of equipment under pressure. If a leak is ever discovered it should not be repaired with tape, but should be taken instead to a qualified repair shop or replaced.

It is a good idea while working never to step on the hose. You should also make certain that the hoses never lie where they can be damaged by hot sparks or slag. It is a good practice when storing welding hoses for a long period of time to put a piece of tape over the fittings to prevent the entrance of dirt or insects. Masking tape works well for this and will ensure that your hoses remain internally clean and clear of obstructions that could damage equipment (FIG. 2-10).

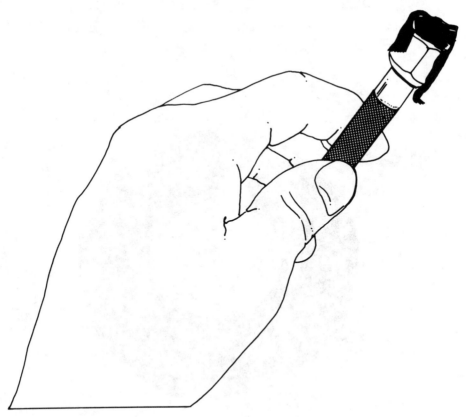

2-10 When storing hoses for a prolonged period, it is a good idea to cover the ends with a piece of tape. This will keep dust, dirt and insects out of the hoses.

TORCHES

The business end of any oxyacetylene welding outfit is the *torch*, or to use the correct terminology, the *blowpipe*. The oxygen and acetylene pass through their respective regulators, through the hoses, and into separate fittings in the handle of the torch. A control knob for each regulates the flow through the torch handle and, in fact, controls the flame (FIG. 2-11).

Currently two types of torches are available: *injector* and *medium pressure*. The major difference between the two is that the injector type can use acetylene at pressures less than 1 psi while the medium pressure torches require that acetylene be supplied in pressures from 1 to 15 psi (FIG. 2-12).

Injector blowpipes

Injector blowpipes, often called low-pressure blowpipes, are all similar in principle but will differ slightly from one manufacturer to another. Generally speaking, they all work as follows. The rear of the torch handle has two hose connections, one for oxygen and the other for acetylene. The front of the handle forms a chamber where the oxygen and acetylene meet for the first time and are mixed before passing through the tip to be burned.

As oxygen passes through its tube in the handle and into the mixing chamber, it creates a suction action that draws acetylene into the mixing chamber as well. Here, the oxygen and acetylene are allowed to expand slightly and to be thoroughly mixed. As the name low-pressure implies, these blowpipes have the ability to work under very low acetylene pressures and are handy for light-duty work such as wire sculpture. They are also commonly used in industry where acetylene is supplied from a generator. Low-pressure blowpipes can also be used with cylinders of oxygen and acetylene (FIG. 2-13).

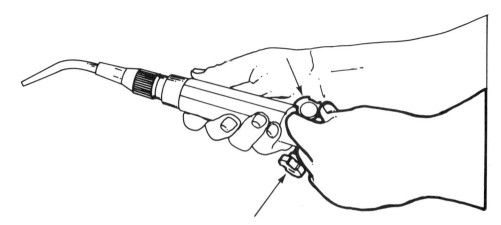

2-11 Control knobs on the torch handle regulate the flow of oxygen and acetylene.

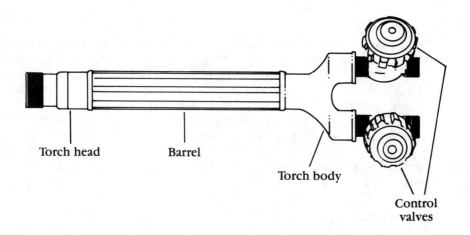

Torch head   Barrel

Torch body

Control
valves

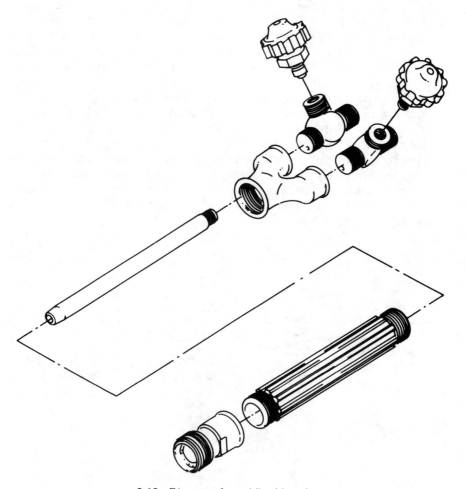

2-12   Diagram of a welding blowpipe.

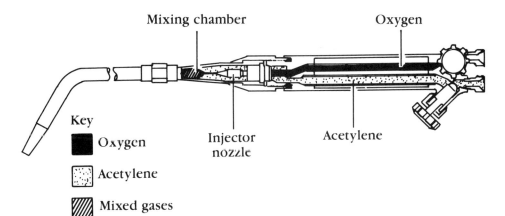

2-13 Internal diagram of low-pressure or injector type blowpipe.

Medium-pressure blowpipes

Medium-pressure blowpipes require acetylene with a minimum working pressure of 1 psi and a maximum of 15 psi. The major difference between a medium-pressure blowpipe, sometimes called a *balanced pressure blowpipe*, and a low-pressure blowpipe is the amount of oxygen and acetylene mixed. Although specific details of this function differ among manufacturers, blowpipes all operate in the same basic manner.

There is a large hole in the center inside a blowpipe surrounded by several smaller holes. These holes are located just behind the mixing chamber and base of the tip. Oxygen passes through the center hole and

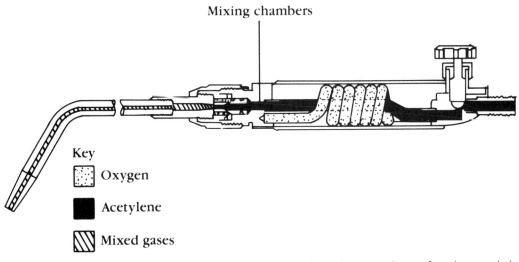

2-14 Internal diagram of medium-pressure blowpipe, sometimes referred to as a balanced pressure blowpipe.

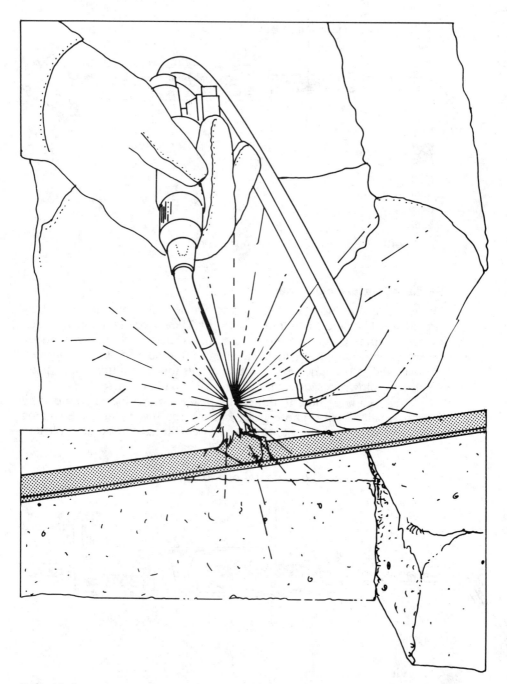

2-15 Medium pressure blowpipes are probably the most common type for general purpose welding.

acetylene exits through the smaller surrounding holes. Some medium-pressure blowpipes are exactly opposite, but the basic principle remains the same. The gases are then allowed to expand momentarily so that thorough mixing can take place before they pass out the tip to be burned.

In all cases, medium-pressure blowpipes are designed to operate at equal pressures of both oxygen and acteylene. Blowpipe manufacturers always supply instructions and recommendations for both oxygen and acetylene pressures so that a specific flame can be developed at the tip (FIG. 2-14).

Medium-pressure blowpipes are probably the most common type of welding torch in use today. They provide as versatile welding tool as the limitations of oxygen and acetylene permit (FIG. 2-15).

TIPS

No welding blowpipe is complete without a welding tip, and there are many different sizes. Torch manufacturers offer a line of various tips designed to be used with their particular unit. Basically, all welding tips are made the same. Most are made of a heavy-gauge copper tubing, which has a high resistance to reflective heat. Some tips are straight while others are curved. In principle, all welding tips are simply a piece of tubing with a threaded end that attaches the tip to the torch body. The other end has a hole through which the oxyacetylene mixture flows and is burned (FIG. 2-16).

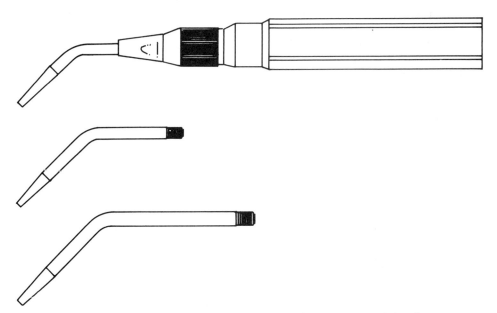

2-16 Different-size welding tips can be used with the same torch handle.

All welding torch makers offer tip size suggestions for the various thicknesses of metals being welded with their particular blowpipe. Basically, the thickness of the metal will determine the tip size. The tip size being used will have a bearing on the pressure settings for both oxygen and acetylene (TABLE 2-1).

Table 2-1   Pressure Chart

Metal Thickness (inches)	Tip Size No.	Size of Welding Rod (inches)	Oxygen (psi)	Acetylene (psi)
1/32	1	1/16	5	5
3/64	2	1/16	5	5
1/16	3	1/16	5	5
3/32	4	3/32	5	5
1/8	5	3/32	5	5
3/16	6	3/32	6	6
1/4	7	1/8	7	7
5/16	8	5/32	8	8

Most welding blowpipes have a cutting attachment that fits on the torch handle in place of a welding tip. Cutting blowpipes (often called the *head* of an oxyacetylene blowpipe) can be either low-pressure or medium-pressure, depending on the type of unit. All cutting blowpipes allow mixed oxygen and acetylene to be released through special preheated orifices in the head of the unit. All cutting tips are the same, though the holes can be of different diameters. A series of small holes surrounds a larger hole in the center of the tip. The oxygen and acetylene mixture burns at the smaller holes. The center hole allows for a stream of pure oxygen to be added to the flame by the welder simply by pressing the long lever on the cutting attachment (FIG. 2-17). It is this flow of pure oxygen that does the actual cutting after the metal has been preheated by the cutting torch flame (FIGS. 2-18 and 2-19).

Cutting attachments

Cutting heads can be of either the injector- or the medium-pressure type. Obviously, only the proper type of cutting attachment should be used with your torch handle. A low-pressure cutting head is attached to a low-pressure blowpipe, for example. The operation of both types is the same as for the respective welding handles covered previously.

Just as there are different-size welding tips, there are also different-size cutting tips. Torch manufacturers supply recommended cutting tip sizes for various thicknesses of metal. Oxygen/acetylene pressures should be adjusted accordingly (TABLE 2-2).

Table 2-2 Pressure Chart for Cutting Tips

Metal Thickness (inches)	Tip Size No.	Oxygen (psi)	Acetylene (psi)
3/8−5/8	0	30−40	5−15
5/8−1	1	35−50	5−15
1−2	2	40−55	5−15
2−3	3	45−60	5−15
3−6	4	50−100	5−15

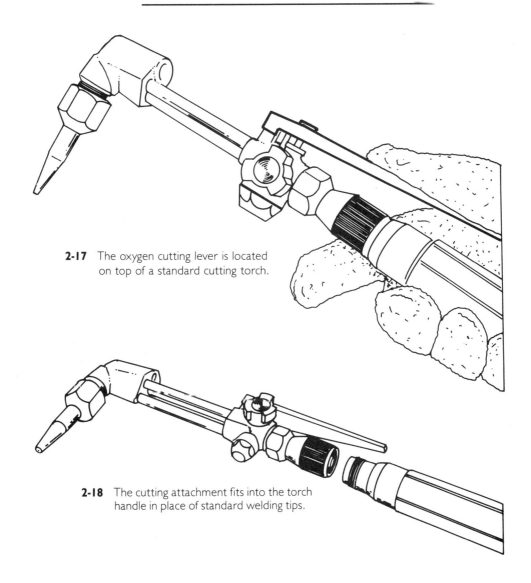

2-17 The oxygen cutting lever is located on top of a standard cutting torch.

2-18 The cutting attachment fits into the torch handle in place of standard welding tips.

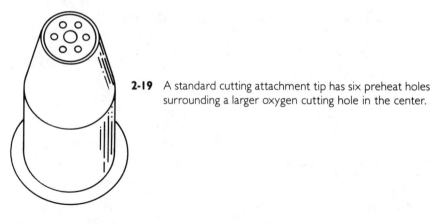

2-19   A standard cutting attachment tip has six preheat holes surrounding a larger oxygen cutting hole in the center.

Slag and flashback

A few things should be done to all tips. Because the cutting or welding tip is necessarily very close to the work, it might become clogged with *slag*, tiny bits of molten metal. When this happens, the flame will sputter and, in extreme cases, possibly go out. The only solution when the flame begins to spit and spurt is to clean the tip with either a special tip-cleaning tool or a suitable-size drill bit. It is important when cleaning any welding or cutting tip, to move the tool in a straight in-and-out motion so the hole is not enlarged or made bell-shaped (FIG. 2-20).

Often, especially when the top of the blowpipe is placed too close to the work, the face of the tip, in addition to becoming clogged, will form a

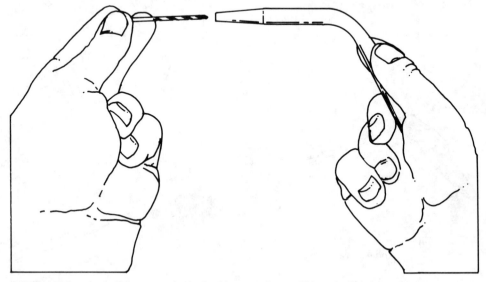

2-20   Welding tips will become clogged with normal use. Although cleaning tools are available, the proper-size twist drill bit works just as well if it is used properly.

buildup of metal. This buildup should be removed with an abrasive—a file or emery cloth. It is important to work carefully to remove carbon and slag deposits whenever they happen (FIG. 2-21).

If a welding or cutting tip should ever become damaged as a result of bending or cross-threading, replace it. A small bend can usually be repaired by simply placing the tip on a block of wood and striking gently with a plastic or leather mallet. Exercise care when doing this so the copper tubing is not compressed (FIG. 2-22).

If a torch tip becomes clogged as a result of slag buildup, the burning oxygen and acetylene might flow back into the hose and regulator, caus-

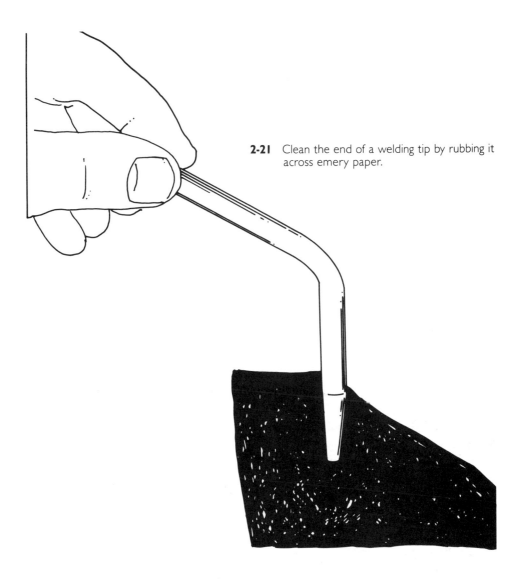

2-21 Clean the end of a welding tip by rubbing it across emery paper.

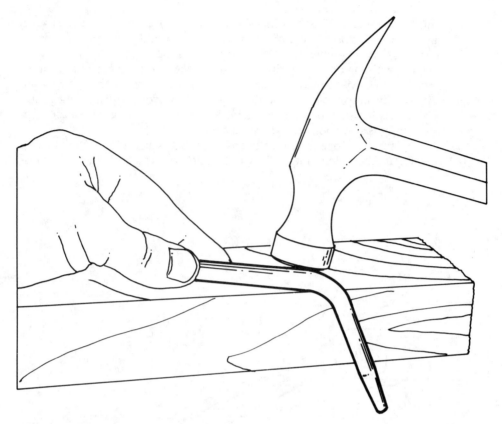

2-22 Straighten bent welding tips with light blows from a hammer. Use a wooden block underneath.

ing what is known as a *flashback*. Special fittings to reduce flashback are available. If your torch does not have them, consider the investment. These fittings are usually attached between the end of the hoses and the torch handle. They will close off the flow of oxygen and/or acetylene whenever there is a flashback and the flow of gas is reversed. Many new welding outfits come with these special fittings (FIG. 2-23).

CLOTHING AND GEAR

In addition to oxygen and acetylene tanks, regulators, hoses, torch, and an assortment of tips, you need some safety clothing and gear, including goggles (with lenses designed for gas welding), leather gloves, a long-sleeved shirt, heavy pants without cuffs, and possibly a hat. Protective clothing is covered in detail in Chapter 14 on safety. Another item the welder will need is some type of ignition system to light the torch. Matches are definitely out of the question.

Probably the most common type of torch lighter is shown in FIG. 2-24. In theory, this lighter produces a hot spark every time the flint is

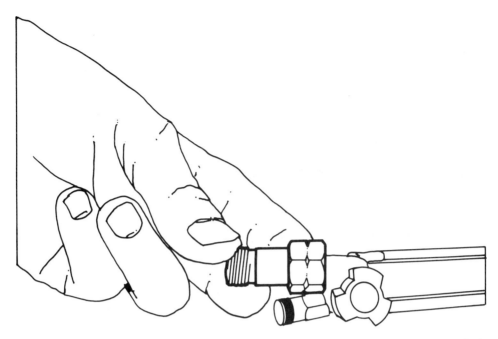

2-23 Flashback arrestors are attached to the hose end of the torch handle and eliminate the possibility of a flashback. Most new torches come with these special safety fittings.

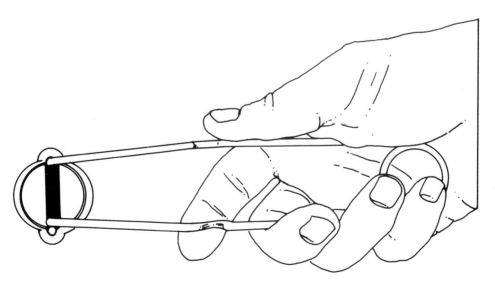

2-24 The standard striker supplied with most welding outfits.

rubbed across the file in the bottom of the cup. In practice, however, this type of torch lighter is not worth the price of the steel from which it is made. Unfortunately, most welding outfits—regulators, hoses, torch cutting heads and tips—usually come with one of these lighters. In my opinion, the first thing you should do when you purchase your welding outfit is to throw out the striker if it looks like the one in FIG. 2-24 and buy a good one instead. Welding supply houses, the same places where you buy your acetylene and oxygen, often carry several different types of strikers.

One relatively new addition to the lighter line is an electronic striker. This type of striker does not rely on the old principle of a piece of flint and steel but instead uses a tiny, hand-operated generator to throw a spark between two terminals. The heart of this little unit is a tiny magneto, similar to the magneto in old-time Model "T" Fords and older tractors. Every time the handle is squeezed, a series of sparks jumps between the terminals. The result is that the torch can be quickly ignited (FIG. 2-25).

WRENCHES

All connections on the tanks, hoses, and blowpipe should be tightened with a suitable-size wrench. All new welding outfits should come equipped with a special wrench for snugging up connections. If your welding gear is older or if the special wrench has been lost, use a substitute open-end wrench. Most experts agree that an adjustable wrench and pliers are poor choices for tightening connections. The main reason is that these tools do not grip as well as wrenches that are made for the purpose. These tools can therefore damage the fittings.

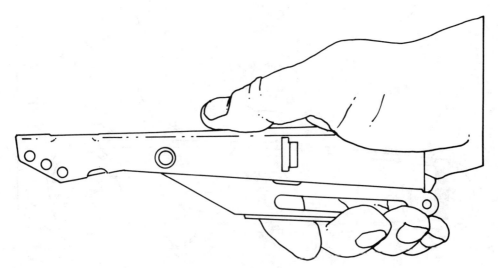

2-25 This new type of torch lighter produces about 10 bright sparks every time the handle is squeezed.

Chapter **3**

Setting up an oxyacetylene outfit

Because the oxyacetylene welding process is the most popular and versatile means of joining metal for the do-it-yourselfer, you should know how to set up the equipment. Some of the basics of torch adjustment and standard safety practices are covered in this chapter. More specific information, such as how to braze, weld, and cut metal, can be found in later chapters.

ATTACHING REGULATORS AND HOSES

To set up for oxyacetylene welding, first make certain that the tanks of oxygen and acetylene are securely fastened to an immovable object, such as a pole or a cart specifically designed to hold the tanks. All welding supply houses carry a line of hand trucks for this purpose. These are a good investment because they safely hold the tanks and give the welder mobility as well (FIG. 3-1).

Next, attach the regulators to the tanks. Crack the oxygen tank. The valve opens slightly so that some oxygen escapes and clears the valve orifice. Wipe the oxygen regulator fitting with a clean, dry cloth and then fit it into the valve on top of the tank. Tighten the regulator fitting then use a special wrench to make sure the connection is tight. After the oxygen regulator has been secured, attach the acetylene regulator to the acetylene cylinder valve in the same manner. Remember that the acetylene fitting is left-handed and must be turned in the opposite direction of the oxygen fitting. As a reminder, note that the acetylene fitting has a notch around the outside of the securing nut (FIGS. 3-2, 3-3 and 3-4).

After the regulators have been attached to the proper cylinders, attach the hoses: the green one to the oxygen tank and the red one with the left-handed threads to the acteylene tank. If the hoses are new, they should first be blown clear with compressed air. Attach the other end of

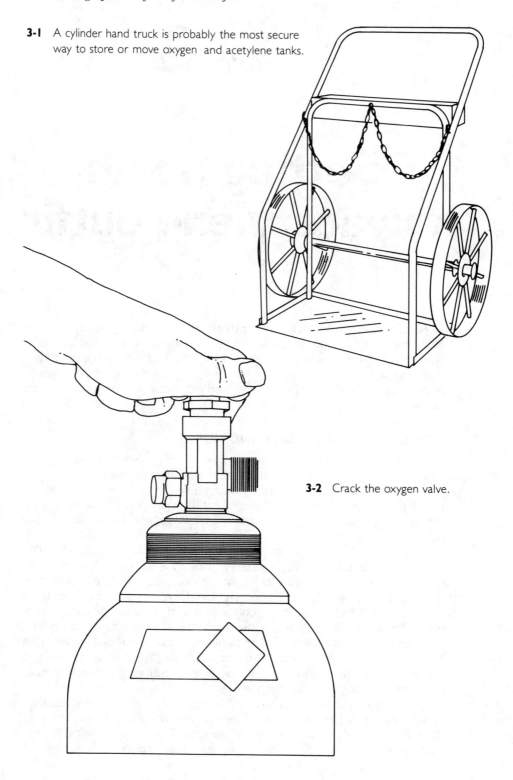

3-1 A cylinder hand truck is probably the most secure way to store or move oxygen and acetylene tanks.

3-2 Crack the oxygen valve.

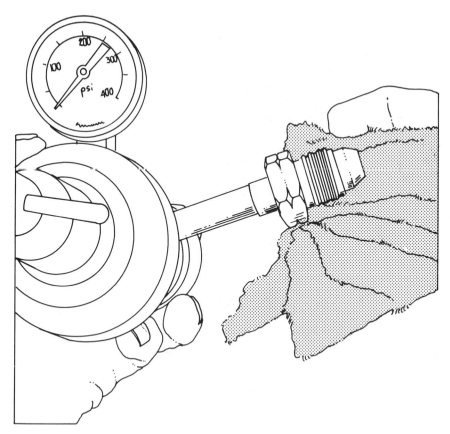

3-3 Wipe the regulator fittings with a clean dry cloth before attaching to the tank valve.

the hoses to the respective fittings on the blowpipe handle. Secure all connections with a wrench. At this point, the valves on both tanks should be turned to the off position, as should the regulators and the controls on the torch handle (FIG. 3-5).

OPENING VALVES

The next step is to very slowly open the oxygen cylinder control valve, allowing oxygen to flow into the regulator. You must do this slowly so there is no surge of pressure on the regulator diaphragm. Once the gauge begins to register pressure, open the oxygen valve one to one and one-half turns, then open the control valve on the acetylene tank. The control valve turns to the left and should also be opened slowly, not more than one full turn (FIG. 3-6).

After all connections are tight and the oxygen and acetylene have been turned on, check that there are no leaks. Because you cannot see

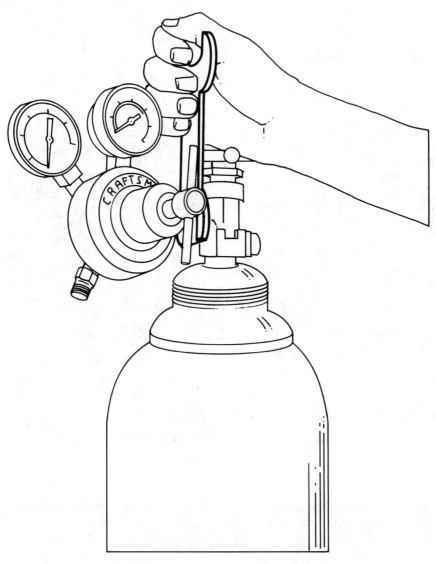

3-4   After the regulator fitting has been snugged up by hand, tighten it securely with a wrench.

leaks, you must check by listening or smelling all the connections. If you are uncertain whether a connection is leaking, use the soapy water test.

SETTING WORKING PRESSURES

Once you are sure all connections are secure, set the working pressures in the hoses. Do this by turning the control levers on the face of the regulators. Consult the chart that came with your welding outfit to determine the proper working pressure for the tip you are using (FIG. 3-7).

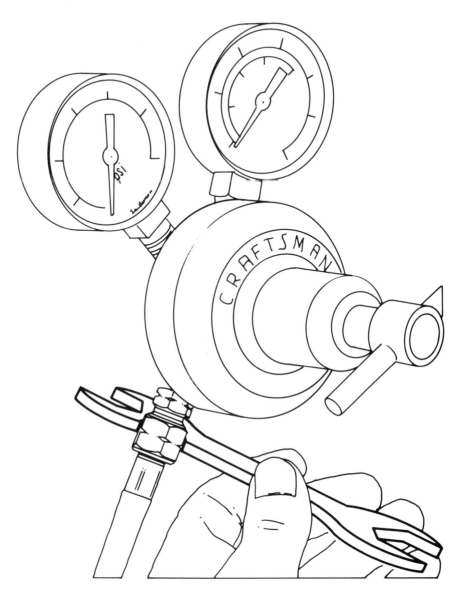

3-5 Tighten hose connections to the regulator with a special wrench.

While setting the working pressures, you might need to open the corresponding control knob on the torch handle. This clears the lines and enables you to make sure oxygen and acetylene are in fact flowing through the handle of the blowpipe. After you have set the working pressures—doing one line at a time—turn off the control knob in the handle and watch the regulator gauges. They should only move slightly, if at all. Set the working pressure for both oxygen and acetylene according to the tip chart for your particular welding equipment.

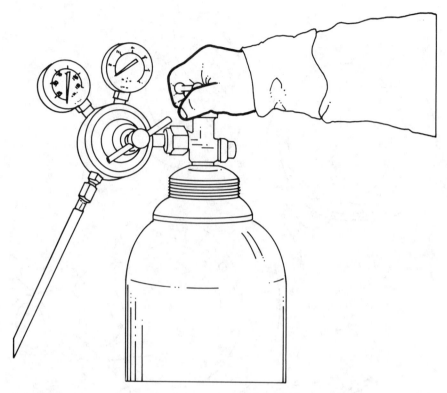

3-6 It is important to open the valve slowly after the regulator has been securely fastened to the tank. This will ensure that there is not a surge of oxygen to the internal parts.

When setting the pressure in the hoses, make certain that there is no open flame or other heat source in the area because gas or oxygen is flowing into the atmosphere. This could cause the fuel to ignite.

LIGHTING THE TORCH

After setting the proper working pressures, light the torch. Before you do, however, you should put on a pair of gloves and goggles. You can raise the goggles to your forehead, so they can be easily pulled down over your eyes when you need them. You should also be standing in front of your workbench, with the hoses coming from behind you so they cannot be hit and damaged by a hot spark or slag. If you are right-handed, hold the blowpipe in this hand and the striker in your left. Turn the acetylene control knob open a crack, just enough to let some gas flow through the tip. Hold the striker close to the tip and squeeze it a few times to generate enough sparks to light the acetylene (FIG. 3-8).

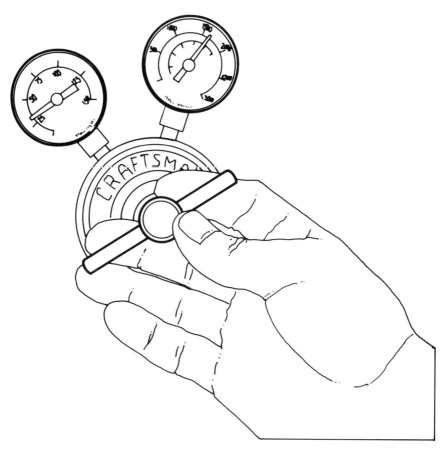

3-7 Set the working pressure in the hose by turning the regulator control lever to the right. The oxygen working pressure is being set at about 30 psi while the pressure in the tank is about 1,700 psi.

Acetylene flame

The first flame that appears will be yellow to orange in color and black smoke will be present at the end of it. This flame is called the *acetylene flame,* and it does not contain any pure oxygen. If too much acetylene is present, the flame will be a few inches away from the hole in the tip, possibly roaring like a jet engine. In extreme cases, this flame will blow itself out. To correct, simply reduce the amount of acetylene flowing through the blowpipe handle. When the acetylene flame is about 8 to 10 inches long, with only a small amount of black smoke, begin to introduce oxygen by slowly opening the oxygen control knob (FIG. 3-9).

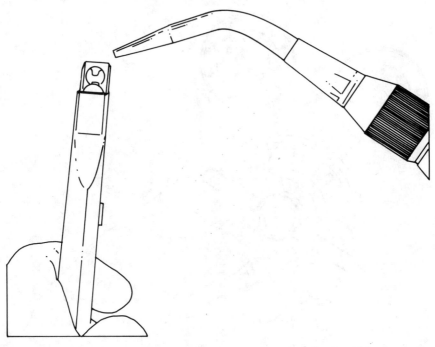

3-8 Hold the striker close to the torch tip, with the acetylene flowing. Light the gas by striking a spark into the gas.

Carburizing flame

As soon as pure oxygen is added to the acetylene flame, changes begin to happen. First, the black smoke totally disappears and the flame changes from a yellow/orange to a whitish color. There will also be three distinct parts to the flame as seen in (FIG. 3-10). The flame produced when oxygen is first introduced is called the *carburizing flame*. This flame, because of the greater amount of acetylene in relation to oxygen, is not enough for welding. It can be used to add carbon to a weld or to blacken the surface of metal.

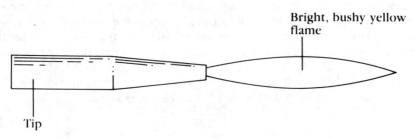

Bright, bushy yellow flame

Tip

3-9 A pure acetylene flame containing no oxygen.

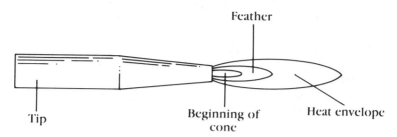

Acetylene

Feather

Tip

Beginning of
cone

Heat envelope

3-10 As soon as oxygen is introduced to the acetylene flame, a cone begins to form and the flame takes on a definite shape. It is called a carburizing flame.

Neutral flame

As you add more oxygen to the flame, simply by turning the oxygen control knob, the flame will change again. The excess acetylene feather will appear to shrink towards the tip and will finally disappear into the small white cone at the base of the tip. This flame is called the *neutral flame* and is used to weld steel and braze most metals. There are only two parts to this flame, a sharp bright cone at the end of the welding tip and a heat envelope (FIG. 3-11). The cone should be well defined and white in color.

The name neutral flame is derived from the fact that the chemical effect of this flame on molten metal during the welding process is neutral. This is true, provided the flame is held in the proper manner, with the inner core not quite touching the molten metal. It is to your advantage to learn how to adjust your blowpipe for a neutral flame, which will be the most frequently used flame in all your welding work. You can adjust the length of the cone by adding more acetylene and then more oxygen. The neutral flame is important not only because it is the most widely used flame; it also serves as a basis for adjusting to or from other flames used in welding and general metalworking.

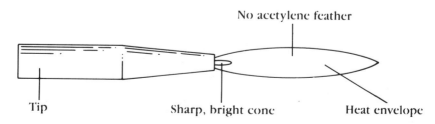

No acetylene feather

Tip

Sharp, bright cone

Heat envelope

3-11 As more oxygen is added, a neutral flame results. This flame is the most commonly used for all welding.

Oxidizing flame

If the oxygen control knob is opened more, the nature of the neutral flame will change again. This new flame is called an *oxidizing flame*. This flame is often used for brazing. To adjust a neutral flame to an oxidizing flame, increase the ratio of oxygen to acetylene. You can either reduce the amount of acetylene coming out of the blowpipe or increase the amount of oxygen (FIG. 3-12).

An oxidizing flame should never be used for welding. The excess oxygen causes breakdowns and defects in both the metal being welded and the weld itself.

The first time you weld with an oxyacetylene outfit, you should practice adjusting the blowpipe to the different types of flames: acetylene, carburizing, neutral, and oxidizing. Once you have become proficient at adjusting to these flames, especially the last two, you can start working on welding technique.

CONCLUDING TASKS

After you have finished welding, turn off your blowpipe. Always turn off the acetylene flow first. Oxygen will continue to flow through the torch, ensuring that the orifice is blown clean. After a few moments, turn off the oxygen.

If you are stopping work for just a short period of time, say 10 minutes or more, close the tank valves, leaving the oxygen and acetylene in their respective hoses. If you are stopping work for a longer period, you should not only close the tank valves, but also bleed the line.

To bleed the hoses in your unit, turn off the flow of oxygen and acetylene into the regulators. Open the acetylene control knob on the blowpipe handle, and let the acetylene in the hose escape. Keep an eye on the gauge of the regulator. As the pressure in the hose drops, the needle will swing to zero. Once this happens, close the control knob on the handle and repeat the operation, bleeding the oxygen line.

Another task you must accomplish when shutting down your welding outfit, after the hoses have been bled, is to release the regulator pressure handles so there is no tension on the diaphragms. Do this simply by turning the handle on the face of the regulator until there is very little or no resistance. This small act does a lot to prolong the life of your regulators.

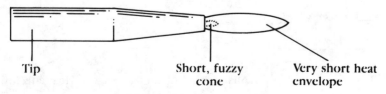

Tip   Short, fuzzy cone   Very short heat envelope

3-12 If more oxygen is added to the neutral flame, the cone will decrease in size. The result is an oxidizing flame.

Chapter **4**

Other welding methods

Up to this point, I have covered only the equipment necessary for oxy-acetylene welding and cutting. The bulk of this book will, in fact, concern itself with this method of joining metals. I believe it is the easiest type of welding for the beginner to learn, and it is versatile enough to enable the do-it-yourselfer to perform most types of metal joining tasks. Remember that all welding falls into three broad categories: gas welding, electric welding, and gas/electric welding. I've already covered the equipment necessary for the first type; now I'll discuss the others.

ARC WELDING

If you feel that you would like to try your hand at *arc welding*, I suggest you contact your local welding supply house for more information and possibly a demonstration. There are also courses offered at vocational schools and adult education classes. In truth, arc welding is not difficult but does require a certain discipline and strict adherence to specific safety practices. Learning the basics of arc welding is best accomplished under the tutelage of a knowledgeable expert.

Shielded metal arc welding, commonly referred to as simply arc welding, is these days probably the most widely used form of welding. It is extremely attractive for commercial work because it is fast and produces strong welds. The major application of arc welding is for joining mile-carbon and low-alloy steels.

Electrodes

The underlying principle of arc welding is quite simple. An electric welding machine produces a specific amount of current that passes through a cable, called an *electrode lead*, to a special handle that holds an electrode. An electrode is similar to a welding rod used for gas welding except that it must also conduct electricity. Electrodes are also covered with *flux*. An arc is struck by moving the electrode close to the metal being welded. The arc is extremely hot, enough to melt both the end of the electrode and the edges of the metal. The current passes through the electrode, melting the end and the metal where the arc from the electrode touches the metal. A work-lead cable connects at the welding machine and is also clamped to the metal being welded, making it possible for the electric current to complete its circular path (FIG. 4-1).

The arc creates a very intense heat, which melts the edges of the metal and the end of the electrode at the same time. The melted metal from the electrode moves across the arc and is deposited on the metal, forming a strong weld, in most cases equal to or stronger than the base metal itself.

When an arc is struck, the flux coating on the electrode produces an inert gas that shields the arc and the weld from the surrounding atmosphere. At the same time deoxidizers are produced that purify the electrode metal. This flux also forms a slag, which protects the molten metal from oxidation. After the weld has cooled a bit, the slag is commonly removed with a special tool (FIG. 4-2).

As you can well imagine, electrodes are a very important part of an arc welding system. They are commonly identifiable by a special numbering system developed by the American Welding Society according to the strength, welding position, usability, and analysis of deposited metal (FIG. 4-3).

Probably hundreds of different flux-coated electrodes have been developed to perform under various applications. Electrodes are available in sizes ranging from 1/16- to 5/16-inch diameter and from 9 to 18 inches long. Fourteen-inch-long electrodes are probably the most widely used in the welding industry.

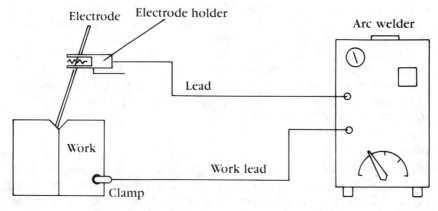

4-1 A simplified diagram of a typical arc welding unit.

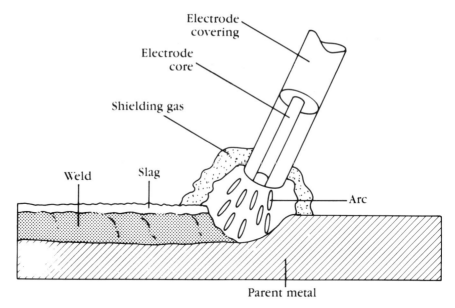

4-2 As the electrode covering melts, a special shielding gas is produced. This gas protects the molten metal from the atmosphere.

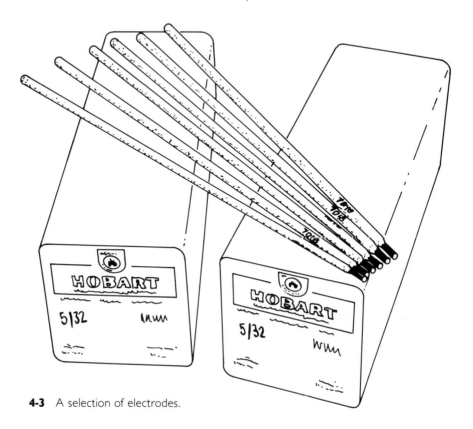

4-3 A selection of electrodes.

Current-supplying machine

Arc welding is made possible by a machine that supplies either direct current (dc) or alternating current (ac). Needless to say, the current must be controllable and adjustable according to needs. At one time, all arc welding was done with dc because ac was unsuitable. Now, however, many welding equipment companies have come out with a line of ac welding units that weld a wide range of thickness and of metal types (FIG. 4-4). One advantage of ac arc welding units is that they can be produced at a lower cost, which makes these units affordable by a larger number of arc weld-

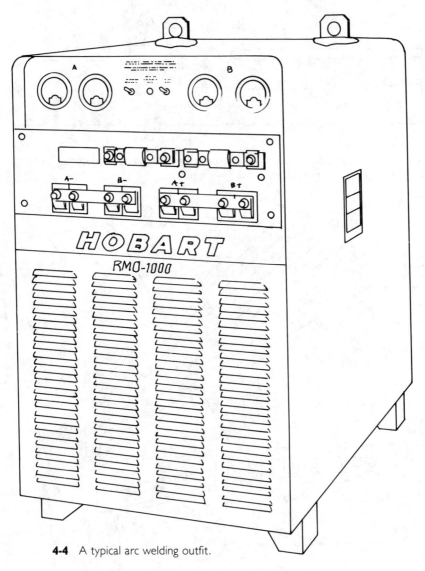

4-4 A typical arc welding outfit.

ers. The ac transformer type is probably the smallest, lightest, and least expensive of all. It is also relatively quiet during operation.

An arc welding machine, either ac or dc, provides the electric power in the proper current voltage to maintain the welding arc. The higher the amperage, the greater the electric flow transferred through the electrode to the metal being welded. Also, the higher the current, the faster the electrode is consumed. Higher settings on welding machines are usually reserved for welding thicker base metals.

Protection gear

Arc welding requires the welder to wear a special helmet to protect the eyes from the extremely bright arc and to protect the face and head from flying molten metal particles. It is also standard practice to wear heavy gauntlet gloves and heavy clothing. The clothing and protection gear is often bulky and confining, but provides a necessary line of defense against burns and eye damage (FIG. 4-5).

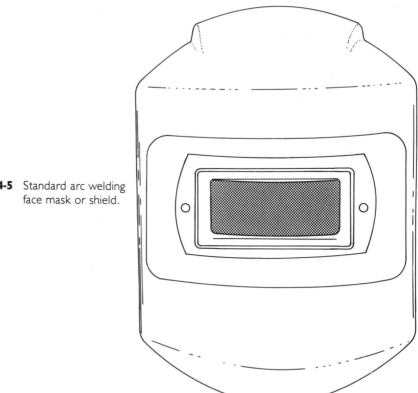

4-5 Standard arc welding face mask or shield.

Striking an arc

While it is not my intention to cover arc welding in detail, I would like to mention some of the basics of this form of metal joining. The first step to be learned is to strike a smooth arc without actually allowing the electrode to touch the metal. When it touches, the electrode sticks to the metal. This is known as *freezing*. If this happens, there are two courses of action, one of which should be initiated immediately.

The first thing you can try when an electrode begins to freeze, is to firmly twist the electrode holder. If a fast twist with an upward pull does not free the electrode, release the electrode from the holder and break the circuit. After the metal and electrode have cooled, you can remove the electrode easily from the metal with a pair of pliers using a twisting motion. Failing this, try striking the base of the electrode with a chipping hammer where it is joined to the metal.

Because of the possibility of freezing and the potential damage to equipment, it is a sound idea to practice striking an arc a few times with the power turned off. Once you have the general idea, you can proceed with the power on.

To strike a proper arc, begin by setting the recommended current on the arc welding machine. Then fasten an electrode in the holder. Hold the electrode, tilted at about a 10-degree angle, almost straight above the metal that is to be welded. If you are using an ac machine, you strike an arc using the scratching method (FIG. 4-6). This is similar to striking a kitchen match. If you are using a dc machine, the accepted method of striking an arc is to tap the electrode to the metal rather than scratch the surface of the metal (FIG. 4-7).

In any case, as the electrode is brought close to the work, an arc will begin. As soon as you see a flash, you should raise the tip of the electrode up from the work until it is about 1/4 inch above the surface and the arc is still present. If the arc is held at this height in a steady position, a puddle of molten metal will begin to form where the arc touches the parent

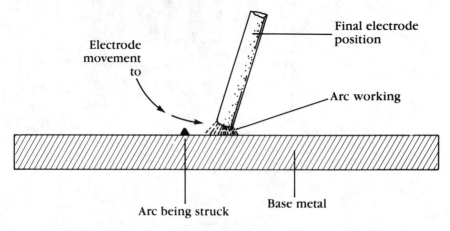

4-6 Use the scratch method to strike an arc with an ac arc welder.

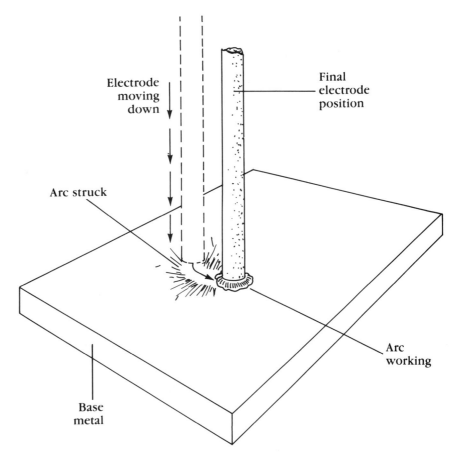

Electrode
moving
down

Final
electrode
position

Arc struck

Arc
working

Base
metal

4-7 Use the tapping method to strike an arc when using a dc arc welder.

metal. At this time the electrode should be lowered toward the metal until it is at a height roughly equal to the thickness of the electrode. For example, if you are using a 1/8-inch-diameter electrode, the tip should be about 1/8 inch above the work. This, incidentally, is about the time that beginners run into a freezing problem as mentioned earlier.

Striking a proper arc should be practiced until an arc can be struck, raised, and then lowered into welding position without creating a short circuit and freezing the electrode. One very real indication that a proper arc has been struck is that the arc will give off a crackling sound, very similar to that of bacon frying. This assumes, of course, that the welding machine has been set properly for a current appropriate for the electrode being used.

Cutting metal

Arc welding equipment can also be used for cutting metal. As with arc welding, arc cutting uses a powerful electric force to melt metal. The molten metal is then either oxidized or blown away with a special attachment that introduces compressed air. Generally, special electrodes are used for arc cutting. These electrodes are coated with a special insulating material that does not conduct electricity. The coating allows the welder to touch the metal being cut without freezing the electrode. The coatings on cutting electrodes act as an arc stabilizer; they concentrate the arc and at the same time intensify the arc's action. Cutting electrodes are nonconsumable, which means that they will retain their original diameters and lengths during cutting.

Arc welding and arc cutting, as mentioned earlier, are not difficult to learn. They should not be attempted, though, without knowledgeable supervision.

TUNGSTEN INERT GAS WELDING

Tungsten inert gas welding, or simply TIG as it is called in the trade, was first introduced in the late 1940s and was an answer to welders' prayers. One of the problems when any two metals are joined with heat is that the molten metal is vulnerable to contamination from oxides in the atmosphere. The end result is a weld that might have weak spots. Before TIG welding was introduced, welders had to resort to various fluxes to remove contaminated material from the weld.

With the introduction of TIG welding, the problem of potentially weak joints and welds was eliminated. Basically, TIG welding is arc welding with the addition of a special inert gas, in most cases argon and/or helium, which shields the area being welded from airborne contaminated material. The gas itself is inert, which means that it has a neutral effect on molten metal. The end result is a very clean, strong weld.

Tungsten inert gas welding differs in another way from standard arc welding in that the electrode used to strike an arc is nonconsumable; it does not decrease in size or length. The heat from this tungsten electrode is sufficient to melt the edges of the metal being welded, but this heat is not strong enough to melt the tungsten. A welding rod, not unlike an oxyacetylene welding rod, is used as a filler material and is not attached to the electrode in any way.

Tungsten inert gas welding was originally developed specifically for welding manganese, aluminum, and stainless steel. Today, it is widely used for nonferrous metals. In the process of developing a new means of welding these metals—which are used extensively in the aircraft industry—the inventors also developed a method that was fast, easy to learn, and more economical than oxyacetylene welding. Tungsten inert gas welding also produces clean, beautifully welded joints.

Unfortunately, the equipment necessary for TIG welding might sometimes be above the means of the casual, do-it-yourself welder. However, the method is widely used in industry and specialty welding shops.

GAS METAL ARC WELDING

To say that the introduction of TIG welding opened the door to new welding methods is an understatement. Unfortunately, TIG welding has its limitations, the major one being that it is not suitable for welding metals with a thickness greater than about $1/4$ inch. *Gas metal arc welding* (MIG) was originally developed as a solution to welding metals thicker than $1/4$ inch by use of the TIG principle of shielding the area being

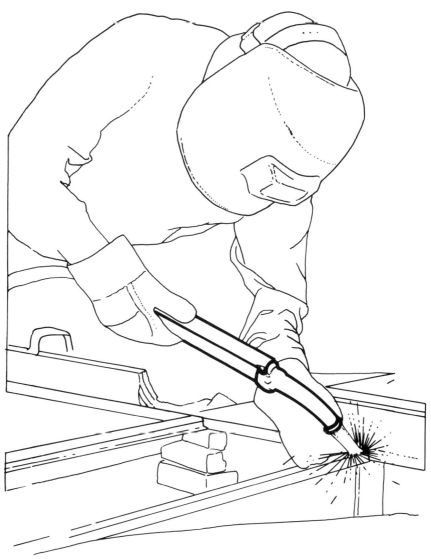

4-8 Gas metal arc welding equipment in use.

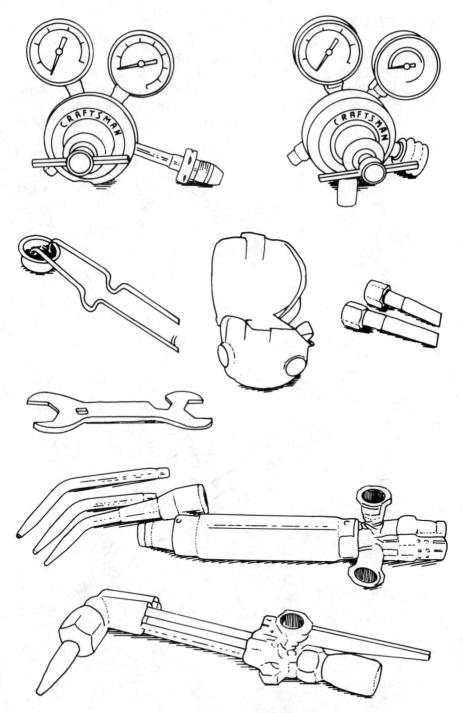

4-9 The typical welding outfit will contain many of the same tools and accessories you will need for repairs and building projects around the home and farm.

welded with an inert gas. Gas metal arc welding, when first introduced, was limited to rather thick welding applications. In a short period of time, certain refinements were introduced to the basic MIG system and transformed this type of welding into a very versatile means of joining metals.

Gas metal arc welding differs from TIG welding in one major way: the electrode is consumed during the welding process. Actually, the electrode in a MIG system is a continuous bare metal filler wire that is constantly fed through the handle of the unit. The wire makes continuous welding possible. This filler wire is fed into the arc and melts on its way to being deposited in the weld. As with TIG welding, MIG welding produces a strong, clean weld. Carbon dioxide is used in MIG welding as the shielding gas. Figure 4-8 shows a MIG welding unit in operation.

Since the introduction of TIG welding in the 1940s and the later development of MIG welding, there have been other additions to the welding field. American know-how has no bounds when it comes to developing a new means of accomplishing a specific task. There are probably another dozen welding processes in use today in addition to the ones covered in this chapter. But these are all used for specific welding tasks in industry and, as with both TIG and MIG, are not really designed for the home welder in either price or application. One possible exception to this, however, is welding equipment that uses a continuous wire for welding.

Wire-feed welding is commonly done with a hand-held gun that contains the end of a length of wire. Electrical current passes through the wire onto the area to be welded. This type of welding falls under the category of arc welding, and it can be either shielded or unshielded depending on the type of wire used.

In all probability, the do-it-yourself welder will learn how to weld by using a basic oxyacetylene welding outfit. The oxyacetylene process can handle a wide range of welding projects—certainly all of those encountered by the casual welder. Oxyacetylene welding and cutting are also a bit safer than other methods such as arc welding, and there is less chance of injury to the user. Another strong point is that the equipment necessary to weld with gas is affordable. There are many welding outfits on the market that can be had for less than $200 (FIG. 4-9).

Chapter **5**

Metals and their properties

Before the do-it-yourself welder can expect to achieve success in welding, he must understand the different properties of various metals. Because metals differ widely in strength, composition, and melting points, it is wrong to think you can join any two pieces of metal simply by clamping them together and heating them up to the melting point until they join. Perhaps a very brief history of metals will contribute to a more thorough understanding of metals.

HISTORY OF METALS

By chance, early man discovered that certain earth soils, when exposed to extreme heat, left a residue after cooling that was very different from stone or the soil. Man later discovered that this hard material could be pounded into various shapes, which were useful for gathering food and making weapons. The actual chemistry of metal was not really understood until later in history.

Metallic ores are oxides, or compounds composed of metal and oxygen atoms. When these ores are heated with a fuel that gives off carbon monoxide—coke or charcoal, for example the ore loses its oxygen as a result of the heat. The end product will be metal and slag. Scientists call this process *reduction*. Because the ore atoms are rearranged and the oxygen removed, the end residue is metal.

Copper

Copper was probably the first metal to be discovered. It is very common and can be heated out of ore with a relatively low temperature. Unfortunately, pure copper is too soft to be used for weapons or tools. If *tin* is added to the molten copper, the cooled metal is much harder. In fact, the amount of tin added gives the metal certain qualities. The addition of tin to pure copper produces an alloy called *bronze*. Man has known about bronze since about 2000 B.C. This metal was first used for weapons, shields, bells, and utensils.

If *zinc* is added to copper instead of tin, the resulting metal is *brass*. There are a number of different brass compositions. Some of these contain iron and tin in addition to zinc. As man learned more about metal, he discovered that bronze and brass did not hold an edge for very long. Eventually, tools and weapons were made from a stronger metal known as *iron*.

Iron

Iron was first discovered as a residue from smelting other metals. Small bits and pieces of iron would be a common by-product in the slag of copper. Eventually, some enterprising individual discovered that these pieces could be heated and then hammered into various shapes. Unfortunately, this early iron turned out to be not as strong as brass or bronze.

It is estimated by anthropologists that around 1400 B.C. the Greeks discovered that heating and reheating iron and then hammering it would transform the iron into something far superior to bronze as far as strength and hardness. It was also discovered that repeated heating of iron caused the metal to absorb carbon from the heating fuel. This excess carbon changes the iron into steel, which is a carbon-iron alloy with a very hard skin. Today, much the same thing is done to obtain case-hardened steel, which is steel with a hard coating and a softer core.

As more experiments were conducted with iron, it was discovered that if red-hot metal was dropped into cold water, the result would be a metal that was harder than any other. Unfortunately, this also produced a metal that was very brittle when subjected to shock. The Romans experimented a bit further and learned that much of the brittleness could be removed from this new metal if it was reheated and then quenched at a higher temperature. The metal retained its hardness but was no longer brittle, making it ideal for weapons. Archaeologists have discovered weapons with these properties dating back to about 1300 B.C.

Steel-making

It was almost 3,000 years ago (about A.D. 1720) before the role of carbon was understood in the steel-making process. A French chemist discovered that if certain amounts of carbon were added to molten iron, the resulting metal would be steel. The problem was determing how much carbon should be added and how the carbon could be combined. Sir Henry Bessemer solved this problem when he invented a steel-making method that is still in use today. The discovery, in 1856, was called the *Bessemer process*. It uses a blast furnace to produce a low-grade steel that is in great demand for various structural steel shapes.

The Bessemer process uses a controlled furnace that blows air through the molten iron. As this air escapes, it takes burnt gases of carbon, manganese, and silicon with it. The end result is a steel that is free of these elements. Then controlled amounts of carbon and/or other alloying elements are added to make a specific type of steel. Alloy steel, which is covered later in the chapter, contains small amounts of tungsten, chromium, nickel, manganese, and vanadium.

Today, steel and alloy steel are made in one of three processes: Bessemer, open-hearth, and the electric arc furnace (FIG. 5-1). The do-it-yourselfer welder should be able to identify metals as closely as possible so he can join the metals. There are three basic areas of characteristics, that the welder should be familiar with: physical, chemical, and mechanical properties. These describe the metal and make it easier to identify.

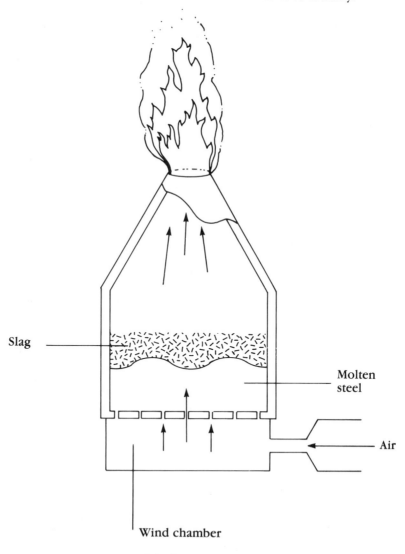

Slag

Molten
steel

Air

Wind chamber

5-1 Bessemer converter.

PHYSICAL PROPERTIES OF METALS

Most of the physical properties of metals can be easily identified by sight. These properties include the *color* of the metal and *magnetism*.

Color

Color describes the surface appearance of a metal. Iron looks different from aluminum, gold is different from silver, copper looks different from zinc, and so on. But there are also similarities. The visual test helps only to identify the most obvious differences among metals. Experienced welders and other people who have worked with metal for a long time can usually come pretty close to guessing what type of metal they are faced with. The rest of us must rely on other methods for identifying metals, however. Nevertheless, the color test is valuable for grouping metals into possible categories.

Magnetism

All ferromagnetic metals—ferrous metals contain iron—have magnetic properties. In other words, a magnet will be attracted to these metals. Actually the magnetic test serves to group all metals into one of two rather broad categories, *ferrous* and *nonferrous* metals. All ferrous metals are derived from iron and its alloys and are magnetic. Nonferrous metals are nonmagnetic and contain either no iron or insignificant amounts. Some examples of nonferrous metals are copper, bronze, brass, zinc, and aluminum (FIG. 5-2).

Melting point

The *melting point* of a metal serves as probably a more accurate identification of specific metals and their alloys. Table 5-1 lists the various melting points of metals and alloys. One very good and accurate way to determine the melting point of a metal, aside from using a particular thermometer, is to use a special crayonlike material that has a known melting point. These melt point indicators are sold at welding supply houses and are available in stick form (very much like a crayon), tablets, and liquids. When these materials are heated, they will melt at a specific temperature—2,800 degrees, for example. When these special melt point indicators are used, the type of metal being considered can be fairly accurately identified (FIG. 5-3).

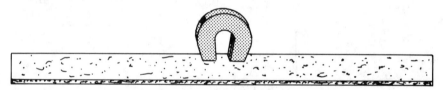

5-2 All ferrous metals are magnetic.

Table 5-1   Melting points
of metals and alloys.

Carbon	3500	
	3400	
	3300	
	3200	
Chromium	3100	
Pure iron	3000	Wrought iron
Mild steel	2900	Stainless steel, 12%
	2800	chromium
	2700	Cobalt
Nickel	2600	Silicon
Stainless steel,	2500	
19% chromium	2400	
Manganese	2300	
	2200	Cast iron
	2100	
	2000	Copper
	1900	
Silver	1800	Brass
	1700	
	1600	Bronze
	1500	
	1400	
Aluminum	1300	Magnesium
	1200	
	1100	
	1000	Aluminum alloys
	900	Magnesium alloys
Zinc	800	
	700	
Lead	600	
	500	
Tin	400	

5-3 A temperature-indicating crayon.

CHEMICAL PROPERTIES OF METALS

The chemical properties of metals include corrosion, oxidation, and reduction. Corrosion is the chemical attack on metal by various elements in the atmosphere and can only happen on aluminum and its alloys. Oxidation in one form is rust and happens on all ferrous metals when certain oxides in or on the metal are combined with oxygen. Some metals, unless protected with a coating of some type, will combine with oxygen in the atmosphere and produce rust.

The ease with which some metals combine with oxygen can complicate the welding process because pure oxygen is used. In certain instances, when oxygen combines with the oxides in metals, reduction takes place, and the volume of the metal becomes smaller. Fluxes and shielding gases (such as those used in TIG and MIG welding) are commonly used to prevent reduction.

The chemical properties of metals can also be determined through certain tests. For example, chromemolybdenum, a popular aircraft steel alloy, can be identified by immersing some filings in a dilute sulfuric acid solution. As the filings dissolve, the solution turns dark green. Chemical tests are not usually used by the do-it-yourselfer, but are used extensively in industry.

MECHANICAL PROPERTIES OF METALS

Undoubtedly the most important characteristics of metals for the welder are their mechanical properties. There is a rather broad category of descriptive terms used to identify the properties of metals: *tensile strength, ductility, hardness, compressive strength, toughness, malleability, impact strength, fatigue, brittleness, stress, elasticity,* and *creep.* Each of these properties is discussed because of their importance to the modern, do-it-yourself welder.

Tensile strength, ductility, and hardness

The tensile strength of a metal is its ability to resist being pulled apart. Imagine pulling a piece of taffy and you get a good idea of tensile strength. The tensile strength of a metal is determined by placing a sample of a metal in a special machine, which is able to pull evenly with tremendous strength, and by noting the point at which the specimen pulls apart. The tensile strength of a metal is expressed or measured in thousands of pounds per square inch (psi) (FIG. 5-4).

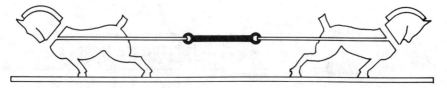

5-4 Tensile strength of a metal indicates the amount of force necessary to pull that metal apart.

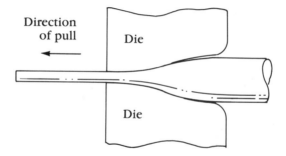

5-5 A metal with good ductility can be worked with force without damage. Wire is made from metal with this quality.

Ductility is a quality of metal that enables it to be formed or worked with force without breaking. In the manufacture of wire, the ductility of the metal is important. Wire is made by pulling metal rod through a special die, which forms the metal without breaking it (FIG. 5-5).

The term *hardness* implies solidity and firmness in a metal. Actually, a metal's hardness is its ability to resist being forcibly penetrated by another material or metal. A number of methods are used to determine the hardness of a metal. Probably the only test that can be done by a home welder is the simple file test. A standard machinist's hand file is used to file the metal. TABLE 5-2 will help measure hardness according to the results of the filing.

It should be noted when performing the file test that several areas of the metal should be tested and an average of the findings determined. Surface defects will, of course, have an effect on the results. Therefore, extremely high or low readings should be disregarded.

Compressive strength, toughness, and malleability

The *compressive strength* of a metal is another mechanical property. A metal is said to have much compressive strength if it is able to withstand a force covering a large area without deforming. It is important that the

Table 5-2 Metal Hardness

Hardness (Brinell Method)	When Filed
100	Metal is very soft, easy to file.
200	Metal reasonably soft, slight pressure required on file.
300	Metal exhibits resistance but can be filed with pressure.
400	Metal is difficult to file; hard.
500	Metal appears almost as hard as file; very difficult to file.
600	Metal cannot be filed; damage to file teeth.

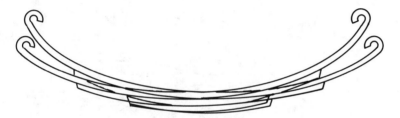

5-6 A metal is tough if it can withstand repeated force without changing. An automobile leaf spring must be made from tough metal.

force be applied slowly and evenly rather than all at once. Striking with a hammer, for example, is not a good test for the compressive strength of a metal. One example of a metal that has good compressive strength is a gasoline engine head or block. This metal can withstand high compression during operation, but it will crack or break if dropped a few feet onto a hard surface.

A metal is said to be tough it it can withstand repeated applications and releases of force without change. An automobile leaf spring is one example of a tough metal. A good one will stand up for many years under the force of a heavy load and return to its original position without change (FIG. 5-6).

The *malleability* of a metal is a metal's ability to be worked cold without noticeable resistance. For example, low carbon steel can be hammered or bent into shape but cast iron cannot. This type of steel, therefore, is malleable while the cast iron is not. Other examples are aluminum, copper, silver, and gold, which can all be cold-rolled into thin sheets without deforming (FIG. 5-7).

Impact strength, fatigue, and brittleness

The *impact strength* of a metal is a desirable metal property in certain applications. A metal with good impact strength must be able to withstand repeated shock without fracturing or breaking down. An example of metals with high impact strength are metals that are used for tools such as hammers and cold chisels (FIG. 5-8).

Metals are said to suffer *fatigue* when they do not hold up under continuous impact or load. Fatigue usually manifests itself in a crack, which lengthens in time. Metal fatigue usually happens when there are flaws in the metal combined with repeated applications of a heavy load. The flaw in the metal gives way because of the excess stress being applied (FIG. 5-9).

The *brittleness* of a metal indicates a lack of ductility. The metal will break, not bend. An example of brittleness is a pencil. It will not bend but will break when a certain amount of force is applied. When a metal is said to be brittle, it will shatter like a piece of glass when subjected to a sudden shock (FIG. 5-10).

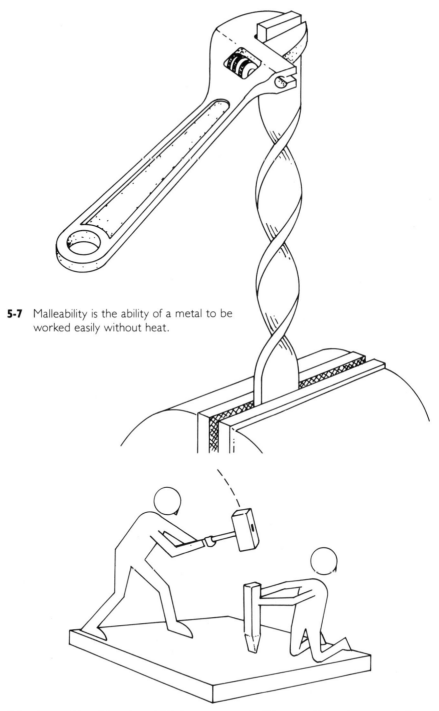

5-7 Malleability is the ability of a metal to be worked easily without heat.

5-8 A metal is said to have high impact strength if it can withstand repeated sharp force. Hammers and chisels are made from metals with high impact strength.

Load

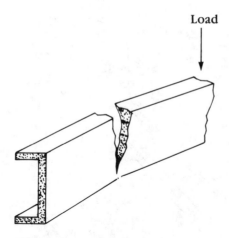

5-9   Fatigue happens when a metal cannot withstand stress from a load.

Stress, elasticity, and creep

Stress refers to the amount of force, such as a load, applied to a metal surface. Some metals will bend when stress is applied. Heat, however, decreases the amount of stress necessary to deform metals. *Strain* is a result of stress.

Elasticity refers to the ability of a metal to return to its original shape after being subjected to stress. All metals have a certain amount of elasticity, as well as an elastic limit. The elastic limit of a metal is that point where it will not return to its original shape (FIG. 5-11).

After the elastic limit of a metal has been reached, any additional force will cause a permanent change in the metal's form. This is known as *creep*.

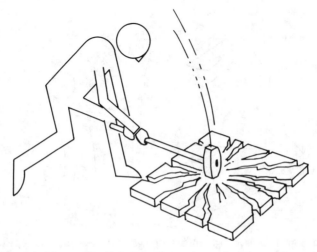

5-10   A brittle metal will shatter when a heavy load is rapidly applied.

5-11 Metal that is elastic can be stretched to a certain limit while it is cold.

IDENTIFYING METALS

Welding methods differ according to the type of metal being repaired or welded. It is, therefore, important to be able to identify metals and their alloys. This is particularly important when you work with various types of cast iron and carbon-content steels. As mentioned earlier, the color or general appearance of a metal can be an indication of the type of metal it is, or, at the very least, the type of metal it is not. See TABLE 5-3. It is to the new welder's advantage to obtain samples of as many different types of metals as possible so he has some basis for metal comparison.

Chip test

One method of identifying different metals is a simple test called the *chip test*. Simply remove a piece of the metal with a cold chisel and hammer (FIG. 5-12) and carefully examine it. While there are obvious limitations to this method, it is still a valuable test that offers fairly accurate results. See TABLE 5-4.

Spark test

Another valuable test to determine the type of metal involves pushing a sample of metal against a bench grinder or grinding wheel and watching the resulting sparks. As you might expect, this is called the *spark test*, a relatively accurate indication of metal type. Various metals give off different spark patterns. With a little experience, almost anyone with a keen eye can identify different types of metals and their alloys. It should be noted, however, that the spark test can only be used for ferrous metals—

Table 5-3 Metal identification by appearance.

	Alloy Steel	Copper	Brass and Bronze	Aluminum and Alloys	Monel Metal	Nickel	Lead
Fracture	Medium gray	Red color	Red to yellow	White	Light gray	Almost white	White; crystalline
Unfinished surface	Dark gray; relatively rough; rolling or forging lines may be noticeable	Various degrees of reddish brown to green due to oxides; smooth	Various shades of green, brown, or yellow due to oxides; smooth	Evidences of mold or rolls; very light gray	Smooth; dark gray	Smooth; dark gray	Smooth; velvety; white to gray
Newly machined	Very smooth; bright gray	Bright copper red color dulls with time.	Red through to whitish yellow; very smooth	Smooth; very white.	Very smooth; light gray	Very smooth; white	Very smooth; white

	White Cast Iron	Gray Cast Iron	Malleable Iron	Wrought Iron	Low-Carbon Steel and Cast Steel	High-Carbon Steel
Fracture	Very fine silvery white crystalline formation;	Dark gray	Dark gray	Bright gray	Bright gray	Very light gray
Unfinished surface	Evidence of sand mold; dull gray	Evidence of sand mold; very dull gray	Evidence of sand mold; dull gray	Light gray; smooth	Dark gray; forging marks may be noticeable; cast-evidences of mold	Dark gray; rolling or forging lines may be noticeable
Newly machined	Rarely machined	Fairly smooth; light gray	Smooth surface; light gray	Very smooth surface; light gray	Very smooth; bright gray	Very smooth; bright gray

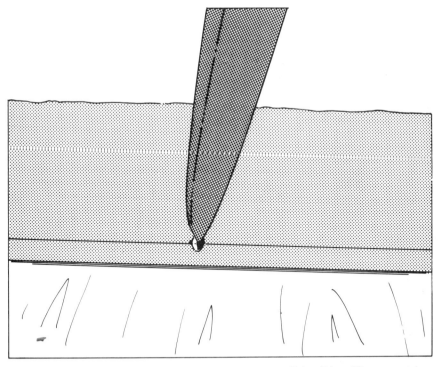

5-12 The chip test is one reasonably accurate means of identifying different metals.

those metals that contain the iron in some form—as nonferrous metals do not have any appreciable spark.

Because the spark test is neither difficult nor beyond the capabilities of the average home welder, it can be helpful to explain how it is performed. A piece of metal is lightly touched against the spinning surface of a grinding wheel. The stream of sparks should then be observed against a dark background and at an angle that permits a good view. Needless to say, you should wear some type of clear eye protection. The length of the stream of sparks is directly related to how much pressure is exerted against the grinding wheel. Three things you should be watchful for are the color, volume, and nature of the sparks (FIG. 5-13). The stream of sparks can then be compared with FIG. 5-14 to determine the type of metal being ground.

Wrought iron is iron with no carbon present. It has a spark stream that is composed of small particles that flow away from the wheel in a straight line. The sparks become wider and brighter as they get farther away from the wheel until finally they go out.

Mild steel contains a small percentage of carbon. It has a stream of sparks similar to wrought iron except for tiny forks in the sparks. The carbon causes the sparks to appear white in color.

Table 5-4 Metal identification by chips.

	Copper	Brass and Bronze	Aluminum and Alloys	Monel Metal	Nickel	Lead
Appearance of chip	Smooth chips; saw edges where cut	Smooth chips; saw edges where cut	Smooth chips; saw edges where cut	Smooth edges	Smooth edges	Any shaped chip can be secured because of softness
Size of chip	Can be continuous if desired	Can be continuous if desired	Can be continuous if desired	Can be continuous if desired	Can be continuous if desired	Can be continuous if desired
Facility of chipping	Very easily cut	Easily cut; more brittle than copper	Very easily cut	Chips easily	Chips easily	Chips so easily it can be cut with penknife

	White Cast Iron	Gray Cast Iron	Malleable Iron	Wrought Iron	Low-Carbon Steel and Cast Steel	High-Carbon Steel
Appearance of chip	Small broken fragments	Small, partially broken chips but possible to chip a fairly smooth groove	Chips do not break short as in cast iron	Smooth edges where cut	Smooth edges where cut	Fine grain fracture; edges lighter in color than low-carbon steel
Size of chip		1/8 in.	1/4 – 3/8 in.	Can be continuous if desired	Can be continuous if desired	Can be continuous if desired
Facility of chipping	Brittleness prevents chipping a path with smooth sides	Not easy to chip because chips break off from base metal	Very tough, therefore harder to chip than cast iron	Soft and easily cut or chipped	Easily cut or chipped	Metal is usually very hard, but can be chipped

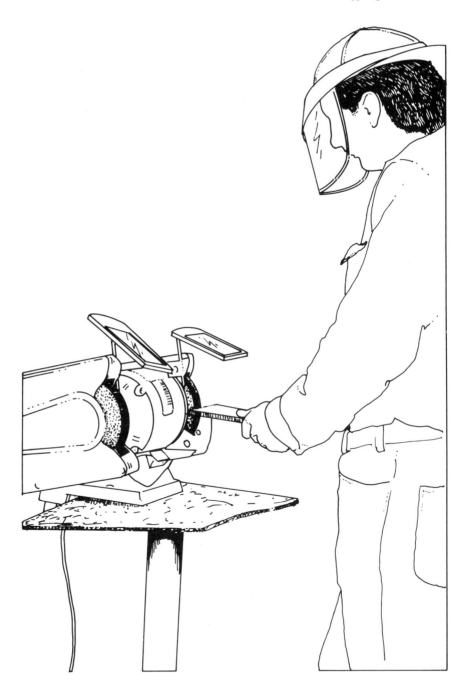

5-13 Use a bench grinder for the spark test.

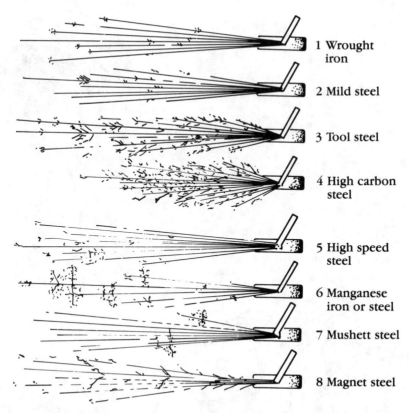

1 Wrought iron

2 Mild steel

3 Tool steel

4 High carbon steel

5 High speed steel

6 Manganese iron or steel

7 Mushett steel

8 Magnet steel

5-14 Representations of the sparks obtained when various metals are lightly pressed against a grinding wheel.

Tool steel produces sparks similar to mild steel sparks but with more forks in the stream. The greater the amount of carbon in the steel, the more forked sparks in the trail. High carbon steel, for example, has a blaze of forked sparks beginning very close to the wheel. Mild carbon steel has sparks with few forks that do not appear until the stream is about 1 foot away from the grinding wheel.

Cast iron has a stream of small repeating sparks that begin very close to the grinding wheel and do not last. The spark stream is very short.

Now that we have discussed the properties of different metals as well as how to determine what type of metal is at hand, it will be helpful to talk about the actual metals and their alloys. The metals that the average home welder will come in contact with are iron, steel and its alloys, copper, aluminum and its alloys, and magnesium. A brief description of these metals follows.

IRON

Iron is extracted from iron ore and is the second most common metal on earth. Aluminum is the most common. Iron is found only in combination with other elements, which must be removed to produce iron. Iron is widely used for such things as bridge parts and castings. Wrought iron was used before steel could be produced economically. Wrought iron is very low in carbon because of how it's made.

Cast iron is wrought iron with the addition of at least 1.7 percent carbon and certain quantities of sulfur, phosphorus, manganese, and silicon. Gray cast iron is cast iron with the addition of *graphite*. Graphite is a form of free carbon that combines with pure iron during smelting and produces a form of iron that is softer than cast iron.

White cast iron is produced when scrap steel and pig iron are melted in an air furnace or *cupola*. This special process produces a white cast iron that is very hard and brittle. Generally, white cast iron is produced along the way to making malleable cast iron.

When white cast iron is annealed, the end product is malleable cast iron. Annealing is a special heating process followed by a very slow cooling, creating an iron that is malleable. Malleable cast iron is a result of a special process that heats up castings (made from white cast iron) and keeps them at a specific controlled temperature. Castings are then cooled during a period of about 24 hours. Although the finished product is not by definition malleable, it possesses great toughness and is not brittle.

STEEL

Steel is an alloy of iron with a carbon content of no more than 1.7 percent. Steel is easily the most common metal for the welder, and there are many different types. By varying the content of carbon or by adding other elements, different types of steel can be produced. Steel with a low carbon content can be soft and easily worked by hand with light tools; steel with a high carbon content, however, will be very hard and might have other characteristics as well.

There is a broad range of carbon steel in use today. Steel can be classified with either a number, such as 440 carbon steel, or simply by designations such as machine steel or tool steel. In any case, these designations indicate the various qualities of a particular steel. For example, machine steel has a low carbon content, 0.10 percent to 0.20 percent and is easily worked by a machine such as a lathe. Tool steels have a higher carbon content, at least 0.60 percent, but lose their hardness at high temperatures because the carbon leaves the metal when heat is applied.

Carbon can be added to molten steel up to 0.83 percent. After this, the tensile strength of the steel is reduced. The higher the carbon content in the steel, the greater the steel's resistance to wear. At the same time, however, strength and toughness are reduced.

Alloy steels are created when other elements are added to plain carbon steel. Basically, an alloy is the combination of any two metals in any proportions. In the case of steel, many different elements and metals can be added to produce a new steel with certain qualities. See TABLE 5-5 for alloying elements of steel.

COPPER

Having been in use for more than 4,000 years, copper has many applications. Probably the greatest use is in plumbing pipes and water lines. Copper is very resistant to corrosion and oxidation. One of the unique properties of copper is that when it cools from a molten state, it expands. This is a favorable reaction in the process of welding pipes and joints. When copper is heated to just below its melting point (about 2,000 degrees Fahrenheit), it becomes as brittle as glass. Copper is a tough (tensile strength of about 20,000 psi) ductile metal that can be easily worked cold or with heat. Copper, is of course, a nonferrous metal that translates into little or no sparks and is nonmagnetic.

ALUMINUM

Aluminum is widely used in industry as well as around the home. Pure aluminum has a tensile strength of about 13,000 psi, which can be increased with alloying and heat treatment to about 80,000 psi. Aluminum can be worked by any method used commercially. Because aluminum is half the weight of iron, it is used extensively in the aircraft industry.

The surface of aluminum oxidizes when it comes in contact with the atmosphere, which can affect welding. This semitransparent oxide film must be removed before welding can succeed. Aluminum poses special problems for the welder and is covered later in this book.

MAGNESIUM

Magnesium is the lightest metal in use today. It equals about one-fifth the weight of copper and less than two-thirds the weight of aluminum. Although magnesium can be worked by all methods, it is subject to rapid oxidation. Therefore, special welding methods must be used. Magnesium is a metal that combines lightness in weight with excellent strength. Magnesium alloys can have a tensile strength of up to about 50,000 psi.

SOURCES OF METALS

Before you can begin to learn how to braze, cut, or weld, you must lay your hands on a supply of metal. Probably the best place to start your search is in the yellow pages. Look under steel, steel fabricators, steel products, metal, salvage, sheet metal, and metal stamping. You can also look under specific headings such as brass, iron, aluminum, and stainless steel, just to name a few. You might also check under steel mills, if you know of any in your area. Metal scrap yards are also a good source of relatively inexpensive steel and other types of metal. The welding supply house where you bought your oxygen and acetylene should be able to tell you where you can purchase steel for welding projects. Local amateur welders are another good source of information. In many cases, because these people make a living welding and repairing metal, they sometimes

Table 5-5   Principal alloying elements of steel.

Element	Melting Point (°C)	Application	Result
Aluminum	658	Little aluminum remains in steel.	Deoxidizes and refines grain. Removes impurities.
Chromium	1615	Stainless steels, tools, and machine parts.	Improves hardness of the steel in small amounts.
Cobalt	1467	High-speed cutting tools.	Adds to cutting property of steel, especially at high temperatures.
Copper	1082	Sheet and plate materials.	Retards rust.
Lead	327	Machinery parts.	Lead and added tin form a rust resistant coating on steels.
Manganese	1245	Bucket teeth. Rails and switches.	Prevents hot shortness by combining with sulfur. Deoxidizes. Increases toughness and abrasion-resistance.
Molybdenum	2535	Machinery parts and tools.	Increases ductility, strength, and shock resistance.
Nickel	1452	Stainless steels. Acid-resistant tools and machinery parts.	In large amounts—resists heat, adds strength, toughness, and stiffness to steel.
Phosphorus (provided by ore)	43	Some low-alloy steels.	Up to 0.05 percent increases yield strength.
Silicon	1420	Precision castings.	Removes the gases from steel. Adds strength.
Sulphur	120	Some machined pieces.	Adds to the steel's machinability.
Tin	232	Cans and pans.	Forms a coating on steel for corrosion-resistance.
Titanium	1800	Used in low-alloy steels.	Cleans and forms carbide.
Tungsten	3400	For magnets and high-speed cutting tools	Helps steel retain hardness and toughness at high temperatures.
Vanadium	1780	Springs, tools, and machine parts.	Helps to increase strength and ductility.
Zinc	420	Wire, pails, and roofing.	Forms a corrosion-resistant coating on steel.
Zirconium	1850	Machine parts and tools.	Deoxidizes, removing oxygen and nitrogen. Creates a fine grain.

sell scrap metal in limited quantities. Professional welders are also a very good source of technical information and advice. It can be to your advantage to establish a relationship with a few.

Obtaining new steel

You should begin welding with new steel. There are fewer problems to deal with than if you were to begin working on rusted scrap steel, for example. This requires a trip to a local steel fabricator, for sheets, plate, tubing, angle, rod, and strap steel. These companies operate by purchasing large quantities of steel from steel mills and keeping inventories for steel workers, who most commonly buy in small amounts.

It has been my experience that people who operate steel fabrication yards are cooperative. Most large companies will often be very helpful in finding the specific metal and gauge or thickness you need. If you are looking for a specific type of metal that is not generally stocked, a certain sheet steel for example, a steel fabricator would be the place to start looking. Keep in mind that special orders for metal usually require a deposit.

Obtaining scrap metal

Once you have become familiar with the various brazing, welding, and cutting techniques using new steel, you will probably want to extend your basic knowledge to other metals, such as high carbon and stainless steels—and bronze. You might also want to try metal sculpture.

Steel fabricators are probably your best source for materials. It is a wise idea to obtain samples of many different types of metals such as chromium, molybdenum, and high-carbon steel so you can widen your experience.

Scrap metal is usually very reasonably priced and commonly sold by the pound. Many interesting shapes, thicknesses, and types of both non-ferrous and ferrous metals can be had with a single trip to the local metal salvage yard. Keep in mind that you will have to devote more time to surface preparation when working with scrap metal. Many sculptors feel the extra effort is worthwhile.

When you visit your local steel fabricator or metal salvage yard, be prepared for the possibility of getting a bit dirty from rummaging through scrap piles. Old clothes, boots, leather gloves, and a hat can protect you from the dirty conditions. A small pocket ruler and a magnet are useful tools to take along as well.

You should also be prepared to carry the metal home. Undoubtedly, the best way is in the back of a pickup truck. If you don't own one, ask a friend who does to go along with you. As a last resort, you can pile the metal on top or inside your automobile. Roof racks and a rope might save a bit of wear and tear on the interior.

Once you get the steel home, you will want to store it so it won't rust. The best storage place is indoors. If you must store the metal outdoors, cover the steel with a plastic or canvas tarp.

Chapter **6**

Welding supplies

*Y*our local welding supply house is your best source for welding supplies and technical information. If you ever have a question about a specific welding project, this is the place to ask. Many of these supply houses also offer instruction and other services. Your welding supply house will be your source for oxygen and acetylene, as well as tools, accessories, fluxes and welding, brazing and special-purpose rods (FIG. 6-1). This chapter focuses on rods and fluxes.

WELDING RODS

During the welding process, it is often necessary to add more metal to the molten area in order to make the weld stronger and to fill in the area. This is accomplished through the use of a special *welding rod*. It is very important that the welding rod match, as closely as possible, the metal being welded (the parent metal) for a strong fusion (FIG. 6-2). During welding the rod melts and deposits its metal into the molten parent metal and contributes to the strength of the weld. The wrong welding rod can result in a weak and ineffective welded joint.

Welding rods are available in a wide variety of sizes and compositions. There is a welding rod for every type of parent metal. The standard length of welding rods is 36 inches except cast iron rod, which is sold at a 24-inch length. Diameters of gas welding rods range from 1/6 to 3/8 inch (FIG. 6-2).

The most common gas welding rods are mild steel, alloy steel, and cast iron. Mild-steel welding rods are usually covered with a copper coating, which protects the rod from rust during storage. Mild-steel welding rods are inexpensive and have a wide application for welding all types of mild steel with a low-carbon content. The tensile strength of mild-steel welding rods is around 52,000 psi.

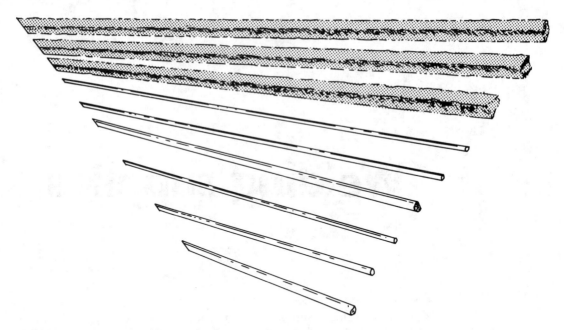

6-1   An assortment of welding and brazing rods. The square rods on the top are for cast-iron welding.

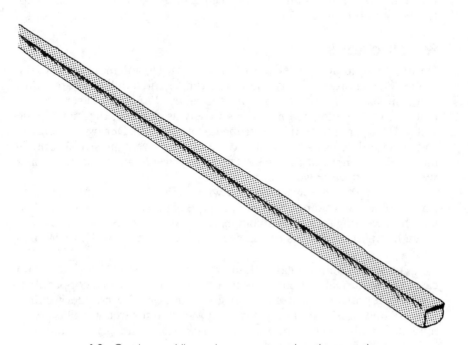

6-2   Cast-iron welding rods are square rather than round.

Alloy-steel welding rods (for oxyacetylene welding) are used for relatively low-carbon alloy steels. These rods are general-purpose welding rods, frequently used for pressure systems such as pipelines. The tensile strength of these rods is about 62,000 psi.

Cast-iron welding rods are available in several different types, depending of course on the application. For example, cast-iron rods are available for welding gray cast iron, molybdenum cast iron (an alloy of cast iron), and straight cast iron. As mentioned, cast-iron rods are available in 24-inch lengths. They are also usually square rather than cylindrical.

In addition to the three basic types of welding rods, there are many other special-purpose rods. These rods are used for joining other types of metals, such as stainless steel and aluminum. Inquire at your local welding supply house for information about the various types of welding rods. You might also want to pick up an assortment of welding rods for experiments and practice.

BRAZING RODS

Brazing rods are used when two metals will be joined by the brazing method. Basically, brazing rods are either bare or flux-coated bronze or brass rods, which are used for brazing or braze-welding steel, cast iron, brass, and bronze. Standard bronze brazing rods have a tensile strength (when used on steel) of about 50,000 psi. The melting point of standard bronze brazing rods is much lower that that of other metals, around 1,600 degrees Fahrenheit.

In addition to the popular bronze brazing rods, other types of rods are available for special metal-joining. These special rods include those made from manganese/bronze, deoxidized copper, and nickel silver. Brazing and all types of rods for metal-joining are covered in detail in chapter 8.

SURFACING RODS

Another type of welding rod is the *hard-surfacing rod*. Certain tools and machinery that are commonly subjected to abrasion, such as a plow blade or the blade on a bulldozer, should be protected against wear by hardening or coating the edge of the tool that receives the punishment. One way to do this is to heat the forward area with a welding torch and then melt a special metal over it (FIG. 6-3). This additional new material is extremely resistant to wear and helps prolong the life of the tool. The most common hard-surfacing materials are nonferrous cobalt-chromium-tungsten alloys. These include nickel-base, iron-base, and tungsten-carbide materials.

FLUXES

Successful metal joining, brazing or welding is directly related to how clean the metals are before they are joined. In some metals, such as aluminum, the formation of oxides during the joining process presents a special problem. To ensure a solid bond, these oxides must be removed from the surface being welded or brazed. In chapter 4, I briefly discussed

6-3   Application of a melted metal to the underside of a plowshare.

shielded arc welding (TIG and MIG welding) and how an inert gas is sprayed over an area during the welding process to prevent contamination. During oxyacetylene welding, there is no such protection, so another means must be used. This protection is accomplished through the use of special fluxes (FIG. 6-4).

A flux, according to the American Welding Society, is a material used to prevent, dissolve, or facilitate the removal of oxides and other undesirable substances from the area being brazed or welded. All fluxes are nonmetallic and fusible. Fluxes cause a chemical reaction to take place resulting in a slag that floats to the top of molten metal during welding or brazing.

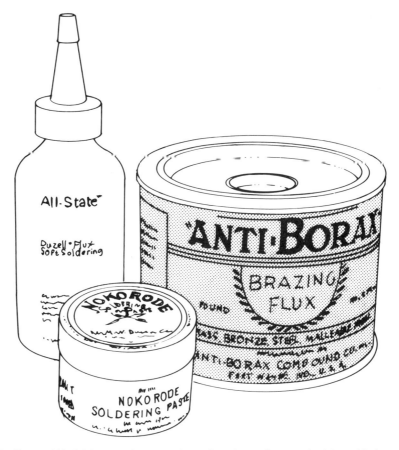

6-4 Fluxes aid in joining certain types of metals and are often required for soldering and brazing.

As you might guess, in order for this chemical reaction to take place, there are fluxes for different types of applications. When welding some types of steel, flux is not usually required because the melting point of the steel is about the same as that of the oxides. Other metals, such as aluminum and iron, do not possess these qualities. Therefore, a flux must be used to lower the melting point of certain oxides so they can be removed and a strong joint result. Special fluxes are always required in the brazing metal.

Fluxes are available for brazing, welding, and soldering. They are sold as a powder, paste, or liquid. Flux-coated brazing and welding rods are also available for special applications. It is always a good idea to check your local welding supply house whenever you are in doubt as to the right type of flux to use for a particular welding project.

Chapter 7

Soldering

*T*here are three ways of joining metal: soldering, brazing, and welding. The strongest, of course, is welding, for which the edges of two or more pieces of metal are heated up to the melting point and fused together. Brazing is similar, except that the metals themselves are not brought up to the molten state but a metal filler material is. Soldering metals produces the weakest joint of all three methods. In many cases, strength is not as important as a complete sealing of the joint surfaces (FIG. 7-1), such as in joining plumbing connections.

The principle of metal soldering is really quite simple. The pieces of metal are joined without bringing either up to the molten stage. Instead, a special metal, solder, is used to bridge the gap between the two pieces. When soldering can be used, it is quite effective. Technically the process is called *adhesion*. In fact, what takes place is that the molecules of solder mingle with the parent metal's molecules and form a strong bond. In some cases, the ingredients in the solder (nonferrous metals) form a surface alloy on one to both of the metals being joined. This creates a solid bond between the metals.

Soldering is similar to brazing in that a nonferrous metal is used to bridge the space between the two metals. The major difference between these two methods, however, is the temperature at which the nonferrous joining metal melts. All soldering is done at a temperature below 800 degrees Fahrenheit. TABLE 7-1 will give you an idea of the temperatures at which soldering, brazing, and welding are accomplished.

TOOLS

Because not as much heat is needed for soldering, different tools are used for it than brazing or welding. Soldering can be accomplished with any number of low heat generating tools such as a soldering iron, soldering

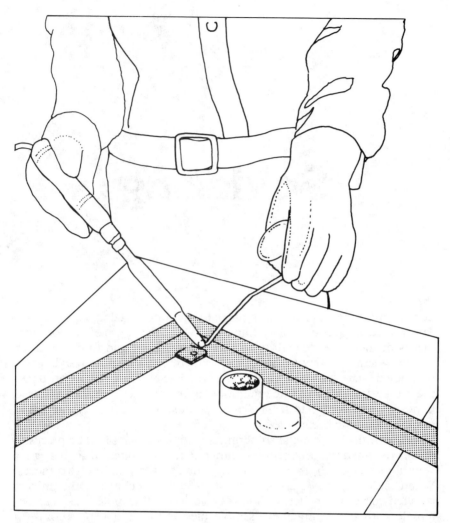

7-1 Soldered joints are never very strong, but in some cases strength is less important than a good seal.

Table 7-1
Temperature
Ranges

Welding	2,800 °F
Brazing	2,000
Soldering	800

gun, or any of the gas (acetylene, propane, natural gas, Mapp gas) torches specifically designed for the purpose. It will be helpful to discuss each of these soldering tools and explain their possible uses.

Soldering irons

Soldering irons are probably the simplest type of soldering tool. A typical soldering iron is simply a piece of steel rod with a wooden handle on one end and a large copper tip on the other end. Sizes can range from small, pencil-like, lightweight types for small electrical soldering projects to large two-handed units used for special industrial applications. The soldering iron must be heated up to a temperature hot enough to melt solder—up to 800 degrees Fahrenheit (FIG. 7-2). This is generally accomplished through the use of a special furnace that is gas-operated (FIG. 7-3).

Other types of soldering irons are more portable because they are heated internally, either electrically or with some type of bottled fuel. These types of soldering irons are illustrated (FIGS. 7-4 and 7-5). The electrically heated soldering iron is best suited for continuous delicate work. The gas-powered unit is limited because of the flame below the tip.

Soldering guns

Soldering guns are quite popular with do-it-yourselfers. They are quick-heating and quite suitable for small repairs and electronic repair or building projects. There are a number of soldering guns on the market, usually in a soldering kit that contains an assortment of tips for different types of soldering (FIGS. 7-6 and 7-7).

Torches

Still another tool used for soldering work is a gas-powered, hand-held torch. There are several variations of this type depending on the intended application. The standard do-it-yourselfer model has many uses around

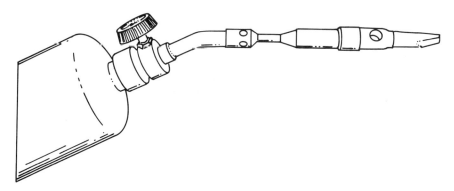

7-2 A portable propane torch with soldering attachment can be used for soldering.

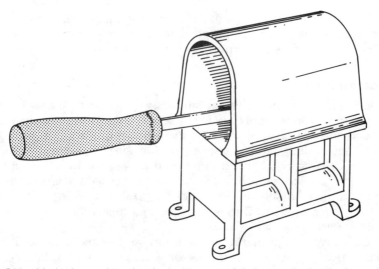

7-3 Old soldering irons, since they had no power of their own, had to be heated in a special furnace, which was most commonly gas-fired.

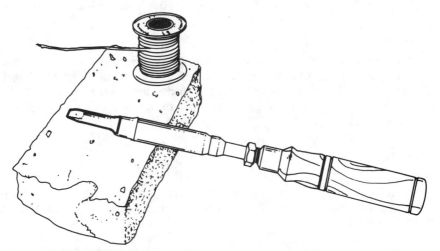

7-4 This soldering iron tip is heated by the flame from a propane torch.

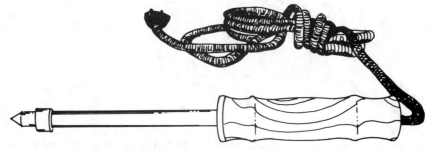

7-5 An electrically heated soldering iron is very handy for continuous soldering.

7-6 A soldering gun kit contains several different-size tips and is handy for a wide range of soldering projects.

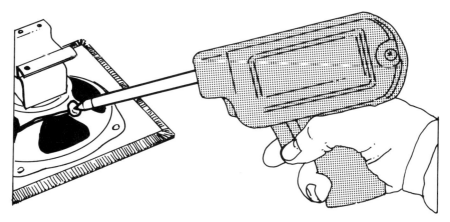

7-7 A hand-held soldering gun is handy for many soldering projects.

the home. The flame itself is used for heating the metal and melting the solder. Needless to say, a flame-powered torch can be used only for certain types of soldering (FIG. 7-8). It is a very handy tool for soldering the joints in a copper-pipe plumbing system. The torch can also be used for thawing frozen water lines, removing paint, and even heating vinyl asbestos floor tile prior to installation. The most common fuel for this type of torch is propane, which is sold in pressurized cylinders that contain enough propane to power the torch for about two hours. Mapp gas is another bottled fuel that can be used in a hand-held torch. Mapp gas is available in standard-size cylinders (usually yellow in color) and offers a much hotter flame than conventional propane.

Propane-powered torch kits are a good investment for the do-it-yourselfer. These kits commonly contain a variety of attachments, such as soldering tips and various-size burner heads (FIG. 7-9).

Professionals who must solder many joints in the course of a day's work will usually have a special torch. These torches, popular with plumbers, have a hand-held torch unit connected by a pressure hose to a regulator that fits into a valve on the top of a 25-pound (or larger) tank of acetylene. These torches can also be run on propane or low-pressure gas and are truly a work-saving tool when used daily. This type of low-heat torch outfit is standard equipment in almost any radiator repair shop (FIG. 7-10).

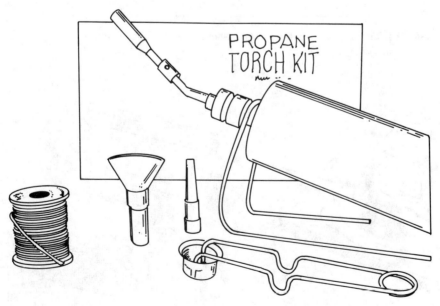

7-8 A good propane torch kit will contain a variety of useful attachments.

7-9 The Spitfire brazing torch has an attachment that aids in soldering plumbing pipes.

7-10 A portable torch outfit is suitable for light cutting, welding, brazing, and soldering.

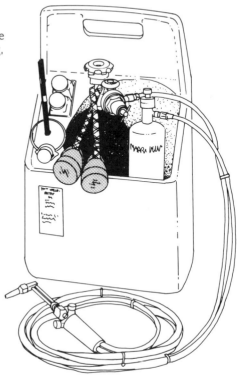

Copper tips

All soldering tools with copper tips—excluding hand-held torches without tips—should provide years of dependable service. Undoubtedly, the most important part of any soldering tool is the tip of the unit. This part is the only one that comes in contact with the work. In order to ensure that this important part is able to do its intended task of heating the metal and melting the solder, there are a few things that you must do.

The tip of a soldering tool should never be allowed to overheat. On electrically powered units this is rarely a problem, as the electrical current is controlled. But soldering irons, which must be heated in a furnace, are often prone to overheating. Overheating a tip results in excessive scaling of the soldering iron core and carbonizing or rosin fluxes. It is difficult to tin an overheated soldering iron tip. One indication that a soldering iron tip is overheating is a greenish tinge that appears when the tip is hotter than it should be. Most authorities recommend that a soldering iron tip never be brought up to a temperature higher than 750 degrees Fahrenheit. In most cases, soldering can be accomplished at lower temperatures anyway, so there is little point in bringing the iron up to this temperature.

Fluxes, which we will discuss in greater detail later, can have a corrosive effect on soldering tips. While fluxes are necessary, you should use the mildest flux possible for the particular job at hand. When you have completed a soldering project, you should clean the tip of flux residue with a damp cloth or steel wool.

If your soldering tool is electrically operated, you should unplug it when not in use. This is particularly important for those soldering tools that remain hot as long as they are plugged in. For soldering guns with a trigger that activates the heat source, this is not quite as important.

You can achieve better results if you keep the soldering tip well coated with molten solder. This is commonly called "tinning the tip," a procedure covered later in this chapter.

Soldering tips will become dirty in time through normal use. For this reason all tips should be cleaned periodically. In most cases a cleaning with a wire brush or steel wool will restore the tip to a like-new condition. Care must be exercised, especially in the use of copper-coated tips, as the coating can be worn off if cleaned with too abrasive a tool. A grinding wheel should never be used for cleaning a tip, for example. After the tip has been cleaned, it should be immediately retinned.

Probably more damage is done to soldering tools through rough handling and dropping than through normal wear and tear. Handle your soldering tools with care, and follow the directions of the manufacturer. Always exercise caution when using a soldering tool, because the possibility of burning yourself always exists. A soldering iron holder offers some insurance against accidental burns and also serves to keep the tool in its proper place. You can usually fashion a soldering tool holder out of a piece of steel welding rod or even a wire coat hanger (FIG. 7-11). You can also rest the hot iron on a brick or other suitable surface.

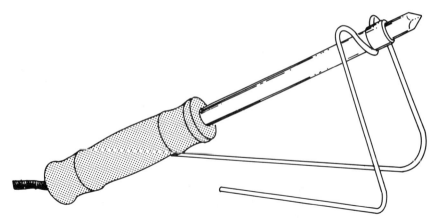

7-11 A soldering iron holder can be made easily from a piece of welding rod or coat hanger.

CHOOSING THE PROPER SOLDER

The key to successful soldering lies in choosing the proper solder for the project at hand. As you might expect, there is quite a large selection of different types of solders on the market. The two most common types of solder used outside the industry, however, are those designated as 50-50 and 60-40. These two solders are the easiest for the do-it-yourselfer to use.

Type 50-50 solder is 50 percent tin and 50 percent lead. The 60-40 solder is 60 percent tin and 40 percent lead. The percentage of tin is always listed first. A chart of the six most commonly used solders is shown in TABLE 7-2.

All solders except the high-temperature 5-95 solder melt at a rather low temperature. You will notice, however, that the more lead in the solder, the higher the temperature must be before the solder will flow or become liquid. This is largely because the greater the amount of tin in a

Table 7-2   Tin-Lead Solders

Number	ASTM % Tin	% Lead	Melting Temperature	Flowing Temperature
5 A	5	95	572°F	596°F
20 A	20	80	361	535
30 A	30	70	361	491
40 A	40	60	361	455
50 A	50	50	361	421
60 A	60	40	361	374

nonferrous alloy, the lower the flowing temperature. The addition of tin lowers the alloy's melting temperature. In an effort to give you a better understanding of these six common solders I would like to explain briefly the uses of each.

5A solder This solder has better quality strength than other types because of its high lead content. It is a good solder for torch soldering where the flame is used rather than a heated tip. This solder can be difficult to work with unless you are quite experienced. For this reason it is generally a poor choice for the do-it-yourselfer. Nevertheless, 5A solder has the ability to provide a very strong joint.

20A solder This type is used extensively in certain industrial soldering tasks such as sealing automobile radiators and auto body work (FIG. 7-12). This is a good solder to use where the finished joint or joist will not be exposed to high temperatures that would cause joint deterioration.

7-12 Most repairs on automotive radiators are made with solder.

30A solder It is similar to 20A solder in that it is widely used for industrial work. This is a common solder for automobile body repair and torch soldering of all types (FIG. 7-13).

40A and 50A solder These are probably the most widely used general purpose solders available today. Both types are used for plumbing pipe joints, roofing seams (where tin or copper is the covering material), and electrical circuit connections such as radio and television. When used for electrical work, the solder usually has a rosin core instead of being solid.

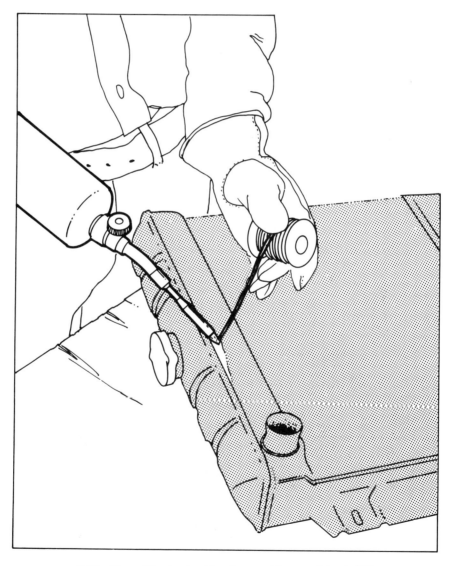

7-13 One of the best solders for repairing a radiator is 20A.

Rosin core simply means that the solder contains a noncorrosive flux that comes from pine trees (FIG. 7-14).

60A solder   This is often called fine solder; it is used wherever temperature requirements are critical, such as in precision instruments.

Another group of solders worth mentioning is called silver solder. This group can contain either lead and silver, tin and silver, or a combination of all three nonferrous metals. Silver solders are used for special joining tasks, such as for certain types of metal or for cases where electrical conductivity is important, particularly in radio or other electrical component circuitry. Silver soldering produces a strong joint. Generally, higher soldering temperatures and special fluxing techniques are called for—not to mention greater skill during soldering.

FLUXES

One of the requirements of successful soldering is that the metal surfaces must be chemically cleaned. They must remain in this state until the solder has hardened (FIG. 7-15). Flux was designed for this purpose. When properly used, it not only cleans the surfaces but seals them from the atmosphere during the soldering. Additionally, some fluxes add alloying elements to the solder or parent metal while other fluxes reduce the surface tension on the metal, and in this way increase the fluidity and adhesion of the solder.

At this point in time, there is no universal flux. Instead, there are several different types of fluxes designed for certain types of metals. Fluxes are available in powder, paste, and liquid form.

7-14   General purpose solders include 40A and 50A.

7-15 Fluxes and solders go hand in hand.

Noncorrosive

All fluxes can be grouped into either of two categories: noncorrosive or corrosive. Generally, noncorrosive fluxes are mild or weak in their fluxing action and are recommended for all electronic and other critical soldering projects. Noncorrosive fluxes are generally suitable for clean copper, brass, bronze, tinplate, cadmium, nickel, and silver. Noncorrosive fluxes leave an unsightly brown stain on the surface. Because this residue is not electrically conductive, it must be removed after the soldering.

Corrosive

Corrosive fluxes are composed of inorganic ingredients. These fluxes are used when a rapid, highly activated fluxing action is desirable, such as in the soldering of cast iron, sheet metal, and copper or brass. Corrosive fluxes generally work well with any of the methods commonly used for soldering.

Corrosive fluxes leave a residue after soldering that can chemically attack the joint just soldered as well as the surrounding metal areas. For this reason, the residue from these fluxes must be removed after soldering. Because of the corrosive action of the flux, corrosive fluxes are

7-16   Acid flux.

unsuitable for certain types of soldering projects, such as sealed containers and most electrical connections. One of the advantages of corrosive fluxes is that they remain relatively stable over a wide range of temperatures, which makes them suitable for the soldering projects requiring higher temperatures.

Probably the most common type of corrosive flux is called *acid flux* (FIG. 7-16). A common formula for this flux consists of 70 percent zinc chloride and 30 percent ammonium chloride. This type is called acid flux because, as the flux is heated and melts, it forms hydrochloric acid. The acid dissolves the oxides and permits the solder metal to adhere to the parent metal. As mentioned, corrosive flux residue must be removed after the soldered area has hardened (FIG. 7-17).

Flux and solder combinations

TABLE 7-3 lists some of the possible flux and solder combinations. Keep in mind that, in some cases, other solder can be substituted, such as 50-50 solder for 60-40. But the best results will be achieved with the combinations listed. It should also be noted that fluxes cannot generally be interchanged. If a noncorrosive flux is called for and a corrosive flux is used, as in the case of electrical wiring, damage might be the outcome of the project. On the other hand, if a corrosive flux is called for and a noncorrosive flux is used, complete bonding might not happen because the flux used was not strong enough to do the job.

SURFACE PREPARATION

The first rule for successful soldering is cleanliness. Everything that comes in contact with the solder—the metals being joined, soldering iron tool tip, and even the solder itself—must be free of dirt, grease, oil, paint,

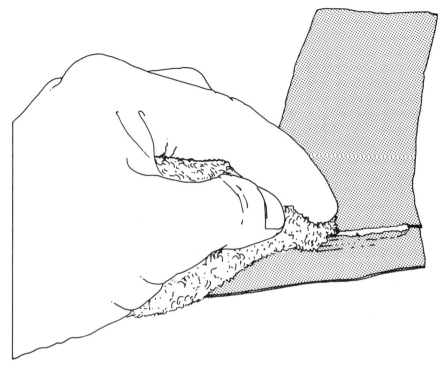

7-17 Corrosive flux should be removed after soldering. A damp cloth will often do the trick. If not, try baking soda.

rust, or corrosion on the surface. Anything else that might interfere with the solder's adhesion should be eliminated.

Of course, the type of project you are soldering will dictate the type of surface preparation. The means of cleaning metal surfaces include steel wool, sandpaper, file, disk sander, grinding wheel, and wire brush (FIG. 7-18). Use any practical method to bring the surfaces up to a bright and shiny condition. Just before you begin soldering, it is a sound practice to

Table 7-3 Suggested flux solder combinations.

Metal	Type*	Flux Solder	Flow Temperature
Tin	N/C	50–50	421 °F
Stainless steel	C	50–50	421
Copper and alloys	C	60–40	374
Iron	C	60–40	374
Galvanized metals	C	50–50	421
Steel	C	40–60	455
Electrical connections	N/C	40–60	455

* N/C Noncorrosive, C Corrosive

7-18 Some of the common tools used in surface preparation.

polish the solder itself lightly with either steel wool or sandpaper. This removes anything on the surface of the solder and ensures that you are using pure solder (FIG. 7-19).

A second rule for successful soldering is that the pieces must fit together well before the actual soldering begins. Don't expect the solder to stand up under a lot of stress and strain because it generally won't. It is far better to have a tight mechanical bond before soldering the work together than it is to expect the solder to do this for you.

In some cases, achieving a sound connection or fit will be fairly sim-

7-19 Clean solder with steel wool before using. This will remove surface dirt.

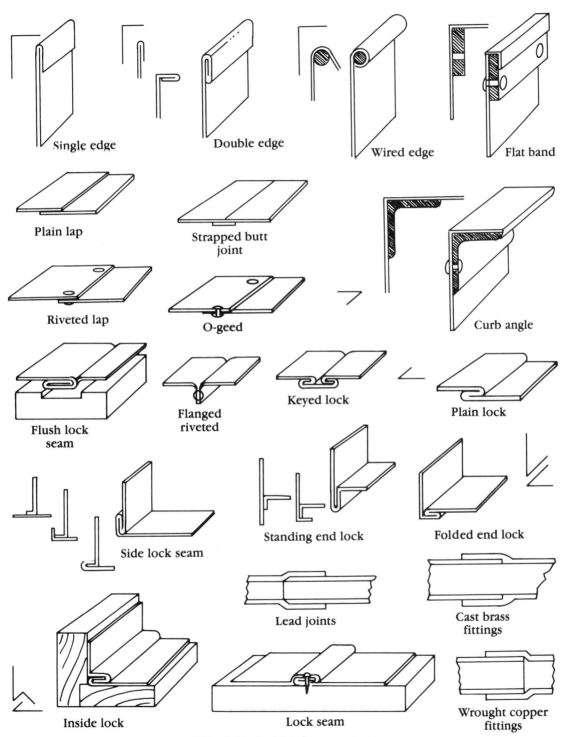

7-20 Soldering joints in common use.

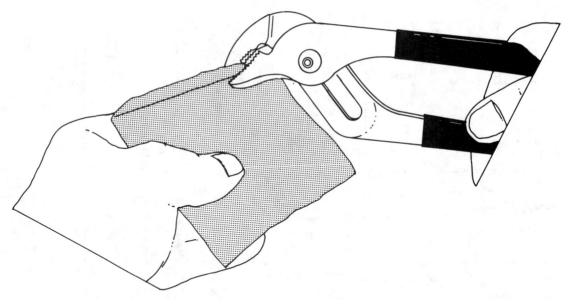

7-21 A pair of pliers is very handy for bending sheet metal to form a soldering joint.

ple, as in fitting plumbing connections together. But, in other cases, you will have to do quite a bit of bending and hammering to get the pieces to fit properly together before soldering. In all cases, a gap of about 0.002 of an inch is acceptable. If the gap is any larger, the solder might not have the ability to hold the work together. The result could be a failing joint.

Whenever possible you should avoid butt joints, as these are the weakest type of soldering. Figure 7-20 illustrates some of the more common metalworking joints that can be used for soldering. As you can see, most of the joints are of the interlocking type. Fashioning this type of joint requires skill that can only be developed through familiarity with the metals being worked and through experience. Nevertheless, there are a few tools that can help you make some of these joints. These tools include a pair of pliers, a bench vise, hammer and anvil, and your own ingenuity. Remember that the better the metals fit together, the better the finished joint will be (FIG. 7-21).

Keep in mind when fashioning joints from sheet metal that this was the way many metals were joined before welding came into existence around the turn of the century. Forming solid joints prior to soldering is a good practice for similar metalworking chores covered in later chapters. In soldering, the better the pieces fit together, the stronger the joints and the better the finished project can be.

TECHNIQUE

Once the proper fit or joint has been established, the piece can be soldered. Before actually beginning, however, it might be necessary to hold the pieces together with a clamp or other suitable tool. This will keep the

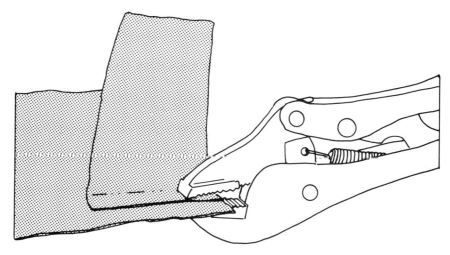

7-22 Vise-Grips are a handy tool to hold pieces secure for soldering.

pieces aligned while you are working on them. It will also be good insurance that the pieces are not disturbed while the solder is cooling, which could result in a weak or ineffective joint (FIG. 7-22).

Tinning the tip

If you will be soldering with an iron or soldering gun, the first task that must be accomplished before the actual soldering can begin is to heat up the tool and tin the tip. Tinning the tip of a soldering tool simply means coating the tip with molten solder. Since it is impossible to solder with an untinned or badly tinned tip—the oxidized film of copper on the surface of the tip prevents the ready transmission of heat—it is important to tin the tip thoroughly. Actually, tinning the tip is a simple process. Begin by heating the soldering tool until it will melt solder (FIG. 7-23). Then simply

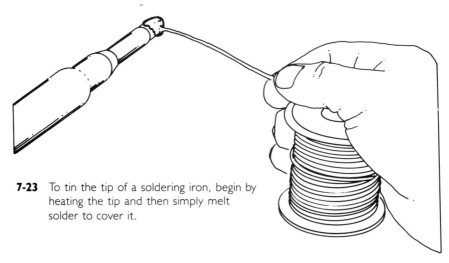

7-23 To tin the tip of a soldering iron, begin by heating the tip and then simply melt solder to cover it.

rub some solder onto the hot tip and spread the molten metal over the surface. When this is properly done, the tip will have a bright, shiny silver appearance.

If you have a problem getting the solder to stick to the tip, it might be because the tip of the tool is dirty. Clean the hot tip with steel wool or a damp cloth; then try again (FIG. 7-24). If the tip is too hot, the copper will tarnish immediately. When this happens, it will be necessary for you to let the tip cool slightly before tinning. You might also find that a little flux rubbed on the tip before applying the solder enables you to tin the tip a bit easier (FIG. 7-25). Use either corrosive or noncorrosive flux, depending on the project at hand.

Heating the work metal

After the tip of the soldering tool has been sufficiently tinned, you can begin the actual soldering. But first apply some flux to the area being soldered. The proper way to solder is to heat up the work with the soldering tool and then apply the solder to the work. The desired technique is to let the heat of the metal melt the solder rather than to melt the solder with the tip of the soldering tool. The obvious reason for this approach is that when the metal is hot enough to melt the solder, you can then expect a reasonably good bond. When the metal is not at the proper temperature, the solder will only lie on the surface rather than bond to it (FIG. 7-26).

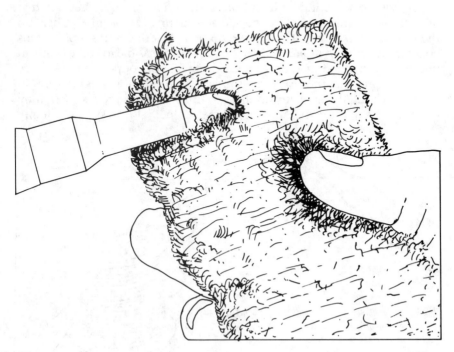

7-24 Clean the copper tip of a soldering iron with steel wool before tinning with solder.

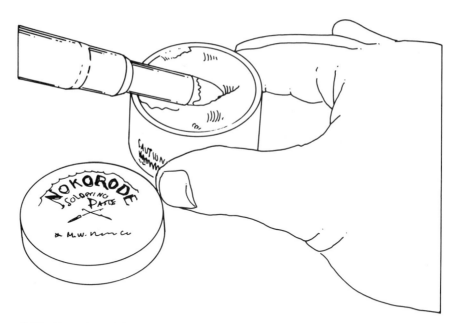

7-25 Sometimes it is necessary to apply flux to the soldering iron tip before tinning.

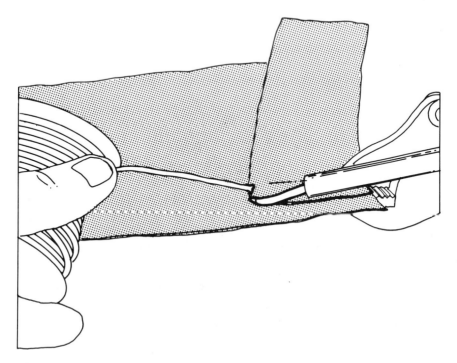

7-26 Always heat up the metal with the soldering iron; then apply the solder to the hot joint. Never do it the other way around.

There are a number of factors that determine how quickly you can heat up the work metal that is to be soldered. These include the thickness of the metal, the size of the tip of the soldering tool—larger tips heat a larger area than small soldering tips—and the power or heating ability of the soldering tool. For small projects that might only require a spot of solder, such as electrical connections, use a very small-tipped soldering tool. For large soldering projects such as a seam in sheet metal, you will find the work much easier to accomplish if you use a soldering tool with a large tip (FIG. 7-27).

Tinning the surface

Once the metal is hot enough to melt the solder, your first objective is to tin the surface. The word *tinning*, in this case, is from an old sheet metal working term. It means to apply a very thin layer or film of solder to the metal surface. Your aim is to lay a thin layer along the entire joint or seam, which will cause the pieces to bond or stick together. Then go back over the work and build up the bead of solder until you have a solid, strong, and even seam.

In some cases, such as installing a radiator bracket, both surfaces are tinned before assembly. The pieces are then pressed together and heated. This causes the mating surfaces, which have already been tinned, to adhere to one another (FIG. 7-28).

As I mentioned earlier, the solder will always flow toward the heat source. You can take advantage of this fact when you need to make solder

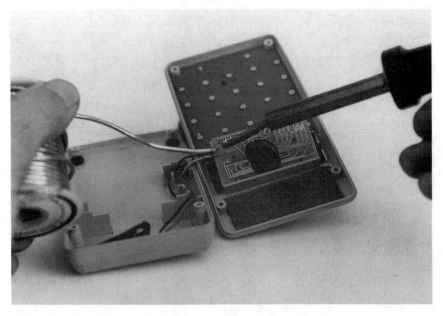

7-27 Small soldering projects are easier to accomplish with a small soldering iron.

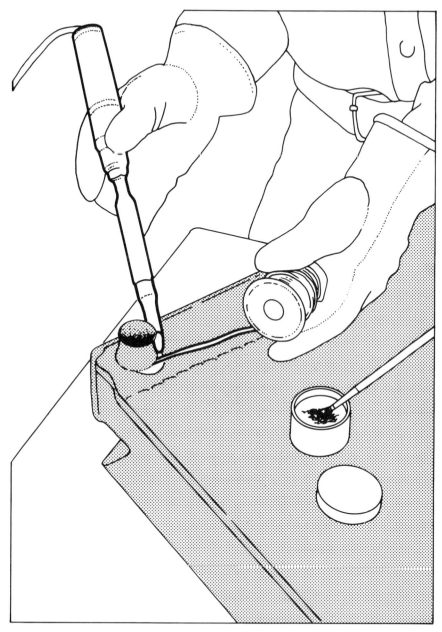

7-28 Some joints are easier to solder if the mating surfaces are tinned with solder and then fitted together and heated until the solder, on both faces, melts together.

flow uphill; for instance, when joining copper tubing for an overhead plumbing system or when soldering a vertical seam. Move the heat source slowly along the seam and the molten solder will follow, leaving a solidified trail.

Chunky solder

When soldering is properly done, the cooling solder appears smooth and slightly shiny. If you are moving too quickly, the solder might take on a grayish cast. This means that the metal was not heated enough to totally flow the solder. In extreme cases, the solder will appear "chunky," with little sharp-edged bits of solder in the bead. This problem indicates that you are working too quickly and that the solder might be the type that requires more heat to flow properly. For example, 5-95 or 20-80 solder requires higher temperatures than 50-50 solder (FIG. 7-29).

Cooling and cleaning the soldered area

After the joint or seam has been soldered to your satisfaction, remove the soldering tool quickly. Remember that the solder will follow the heat. Once you have removed the heat source, look over the work carefully. If any sharp points are present, or any areas appear not to be covered sufficiently, touch up with the tip of the soldering tool. When you are satisfied with the joint or seam, let the work sit undisturbed until it has cooled. Moving the still hot metal might cause the seam to come apart, and this is the reason for not touching it until it has cooled.

 Most soldering involving a flux also requires that you clean the just-soldered area after it has cooled. In the case of noncorrosive flux, there is no danger that the flux will damage the joint or surrounding area. There will, however, be a brown residue around the soldered area, especially if rosin-core solder is used. This brown film is very brittle, actually a pine-tar residue and can be removed quite simply with the tip of a pocketknife or other suitable tool (FIG. 7-30).

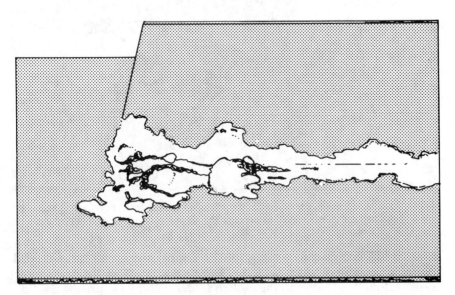

7-29 A solder joint that appears "chunky" is usually a case of not enough preheat.

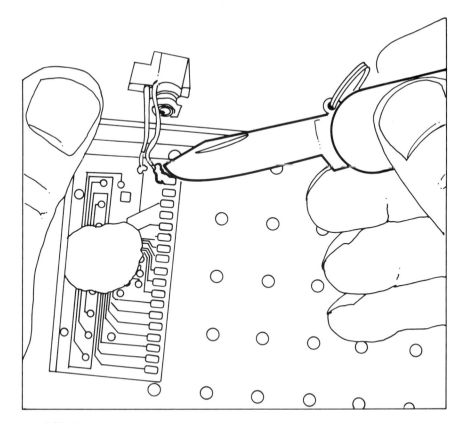

7-30   You can usually remove noncorrosive flux with the tip of a pocketknife.

Corrosive flux must be removed from the finished soldering project, as stated earlier, because the corrosive flux can damage the surrounding metal. This is usually best accomplished by rinsing the area with water. If this is not possible, use a damp cloth (FIG. 7-31).

Soldering with a flame

When soldering with a flame, instead of a soldering tool with a copper tip, the same basic procedures are followed:

- Cleaning the surfaces.
- Fitting the pieces together.
- Applying flux.
- Heating the metal until it is hot enough to melt the solder.
- Leaving the metal cool undisturbed.
- Cleaning the soldered surface of flux and smoothing if necessary.

Generally, soldering with a torch requires a light touch with the flame and strict attention to the work. The flame can be hotter than 800 degrees

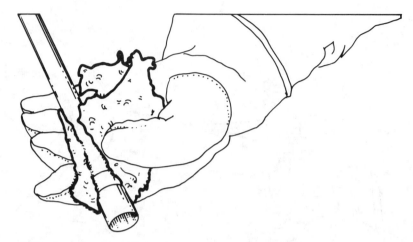

7-31 Remove corrosive flux from a cooled soldered joint with a damp cloth.

and the possibility always exists of overheating the metal. This can result in poor adhesion of the solder, scorching the metal, and in extreme cases, warping the metal.

Soldering copper plumbing pipe

Probably the most common use of a torch in soldering is for making soldered joints in a copper-pipe water line system. It will therefore be helpful to explain how this is done.

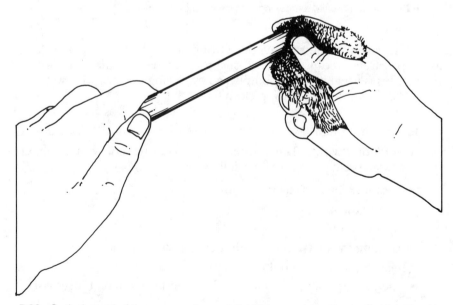

7-32 Both the end of the copper pipe and the fitting must be clean and shiny. Do this with steel wool.

Begin by cleaning the end section of pipe with either fine emery cloth or steel wool (FIG. 7-32). The same treatment is given to the inside of the copper fitting. All mating surfaces should be bright and shiny. Next, apply the flux (FIG. 7-33). In most cases, this is a corrosive paste-type flux, a standard in the plumbing trade. The pieces should then be fitted together tightly and positioned so they will not move during the soldering process or until the solder has cooled and solidified totally.

Apply the heat to the pipe and fitting, moving the torch to ensure that the parts are being evenly heated (FIG. 7-34). As the flux heats up, it will begin to bubble and possibly turn a brownish color. This is usually an indication that the copper is at about the right temperature to melt the solder. Move the flame of the torch away from the joint and touch the end of a strip of solder to the joint area (FIG. 7-35). If the metal is hot enough, the solder will flow into the joint (FIGS. 7-34 and 7-35).

When soldering copper water lines, it is important that you do not apply too much solder. Actually you do not need very much. Only a small

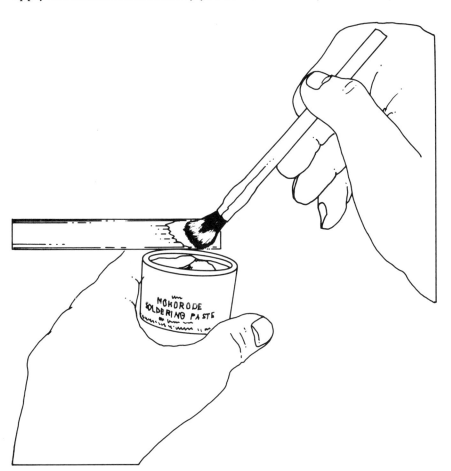

7-33 Apply flux to the clean pipe and fitting with a small brush.

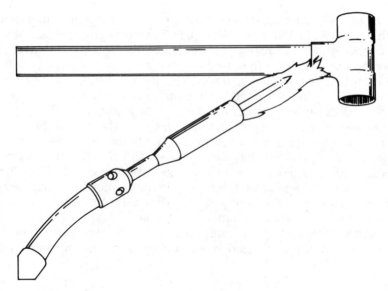

7-34   The fluxed fitting is twisted onto the pipe; then heat is applied with a propane torch.

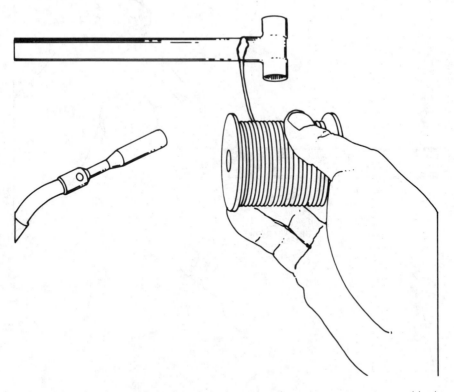

7-35   After the pipe and fitting have been heated to the proper temperature, solder is touched to the hot joint and will flow into place, forming a tight seal.

amount of solder should be visible around the outside of the joint. Excess solder buildup around the fitting and pipe can actually weaken the joint rather than make it stronger.

After the proper amount of solder has been applied to the joint, wipe the area with a damp cloth. This gives the joint a nice finished appearance, as well as cleaning the area of flux. The wiping will also remove excess solder (FIG. 7-36).

SAFETY CONSIDERATIONS

Soldering is a relatively safe method of joining metals, but as with any operation that involves heat, there is always the possibility of burns, noxious fumes, and damage to the skin. In most cases, accidents can be avoided by being careful while working. It is a good idea to have a specific place for the soldering tool when it is turned on but not actually in use. This can be a wire holder as shown in FIG. 7-12, a large ashtray (although these tend to have a rather short life in the workshop), or even a

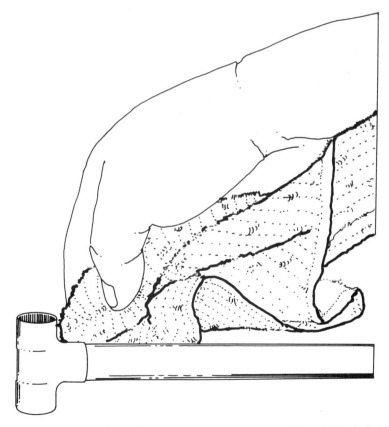

7-36 Before the solder has totally hardened, wipe the joint with a damp cloth. Work carefully, or you might disturb the joint.

brick. If the soldering iron or gun is electrically operated, as most types are, you should set up your work area so there is no possibility of burning the power cord with the tip of the unit. This can cause a short circuit or even electrical shock.

Eye protection

When you are working with corrosive fluxes, it is a good idea to wear a long-sleeved shirt and eye protection. In some cases a pair of gloves might also be a good idea.

The chances of mishap increase when you find that you must work away from a workbench. This is usually the case when working on a copper water line system. Precautions should be taken. If working overhead, wear eye protection, a long-sleeved shirt, and a hat.

Ventilation

Whenever you are soldering with a gas-powered torch (for example, a propane torch), you should make certain that there is adequate ventilation in the work area. If you are soldering copper pipes, keep an eye on combustible areas around the work, such as wooden flooring, joists, and wall studs. In many cases, you will be forced to solder very close to these areas. The possibility of fire is always present so have some means of putting out a fire if one should start. Some good fire protection or fighting tools to have around are a bucket of water, a fire extinguisher, a box of baking soda, or a plant atomizer. The last item will quickly spray a cloud of water on a smoldering or burning piece of lumber and quickly put out the flame.

If you find it necessary to solder close to wooden joists, slip a piece of asbestos board between the work and the joist. The asbestos will prevent the possibility of a fire.

Since all fluxes give off fumes or smoke when heat is applied, ventilation is more than important. While noncorrosive rosin fluxes are not harmful, fumes can upset your system for a short period of time. It is therefore not a good idea to breathe these fumes for long periods. Prolonged inhalation of corrosive flux fumes should be avoided. If the soldering project is a large one, some type of respirator should be worn to protect the worker.

Soldering is an effective and enjoyable means of joining metal, and it is also rather inexpensive. It is good practice for the stronger methods of joining metals because some of the techniques are basically the same. Approach all soldering tasks with caution, and develop safe work habits. Not only will this help you achieve predictable results, but it will help prepare you for other methods of joining metal.

Chapter **8**

Brazing and
braze-welding

*I*n chapter 7, I covered how to join metals by using low heat and a non-ferrous filler material. Recall that soldering takes place at temperatures below 800 degrees Fahrenheit and produces the weakest joint of the three types of metal-joining techniques: soldering, brazing, and welding. In this chapter, I'll cover how to join metals with a nonferrous filler material; only in this case, at temperatures higher than 800 degrees. Before I discuss metal joining, it would be helpful to explain and define two often-confused terms: brazing and braze-welding.

USES OF BRAZING

Brazing and braze-welding are two different techniques for joining metals. Brazing refers to a process where the metal parts are heated to temperatures higher than 800 degrees Fahrenheit (FIGS. 8-1 and 8-2). A nonferrous filler metal is then introduced into the joint. Part of the key to successful brazing is that the filler metal must have a melting point lower than that of the parent metals. The pieces to be joined are heated to a point where the brazing filler metal will melt, but not the parent metals.

When you are brazing metals, the fit between the pieces is very important and must be tight. Since the filler material flows freely over the surface and into the recesses of the joint, it will not fill gaps or other areas that do not fit together well. This flowing of the filler material is known as *capillary attraction*. It happens when the contacting surfaces of a liquid (the molten filler material) and a solid (the parent metals) interact so that the liquid surface is no longer flat. A coating of flux aids the capillary attraction and helps the filler adhere to the parent metals.

A brazed joint is stronger than a soldered joint and, in certain instances, is as strong as a welded joint (where the parent metals have been heated up to the melting point and have fused together). Brazing is

8-1   Brazing in progress.

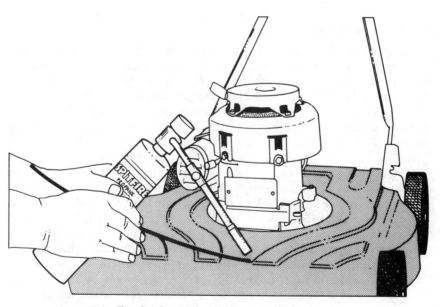

8-2   The Spitfire torch can be used for light brazing.

used where mechanical strength and pressure-proof joints are required, such as in a pipeline system and cast transmission cases (FIG. 8-2).

Because of the lower temperatures used in brazing, metals being worked on will not warp. Brazing is also much quicker and less expensive than welding because the metals need not be heated to the melting point.

Brazing is a suitable means of joining many types of metals, such as steel, cast iron, nickel, copper, brass, bronze, cast bronze, magnesium, aluminum, monel, and inconel. To successfully join these metals, it is necessary to fit the pieces together carefully. Use the proper flux and filler material and work at a predetermined temperature. Brazing is also a good technique for joining dissimilar metals such as nickel alloys and copper alloy base metals.

USES OF BRAZE-WELDING

Braze-welding differs from brazing in the design of the joints being secured. Braze-welding is used for groove, fillet, and other weld joints where the filler material does not take advantage of capillary action. Braze-welding does, however, use a nonferrous filler metal that has a melting point less than the parent metal but higher than 800 degrees Fahrenheit.

The technique for braze-welding is similar to fusion welding. The important exception is that the parent metal is not melted but only raised to the tinning temperature.

FILLER MATERIAL

Both brazing and braze-welding use a nonferrous filler material. In most cases, an alloy of copper and zinc or copper and tin. The two most commonly used copper alloys for brazing and braze-welding are brass and bronze.

Brass is an alloy consisting of copper and zinc, while bronze is comprised mainly of copper and tin. The more popular for both joining techniques is brass. Note that even experienced welders call brazing and braze-welding rods *bronze* when actually they are using brass rods consisting of copper, zinc, and about 1 percent tin (FIG. 8-3).

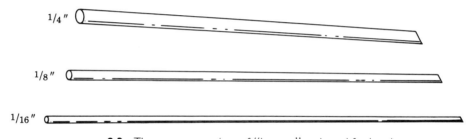

8-3 Three common sizes of ''bronze'' rod used for brazing.

In addition to the copper alloys, there are other brazing and braze-welding rods that are used for joining special metals. These include nickel and chromium alloys, silver alloys, copper and gold alloys, aluminum and silicon alloys, and magnesium alloys. It is probably a safe assumption that the do-it-yourself welder will use bronze rod for most types of brazing and braze-welding. This is a good nonferrous filler material for joining cast iron and steel. For joining other types of metals, it is always a good idea to consult your local welding supply house for the best type of nonferrous filler material.

FLUX

Flux plays a very important role in braze-welding and brazing. Most brazing processes require the use of flux to prevent oxidation during the application of heat. A flux, you might remember, is used in soldering to protect the metal surfaces that are being joined from the atmosphere. In brazing and braze-welding the flux is used for the same reason. In addition, a flux will aid the capillary attraction of the nonferrous metal during the brazing process, and at the same time will dissolve the oxides that form during either brazing or braze-welding.

Flux is commonly available in powder, paste, and liquid forms. Also available are bronze rods that have been coated with a flux, which eliminates the need for additional flux. There is no such thing as an all-purpose flux. There are, instead, fluxes designed for specific applications. Most fluxes are chemically pure. Read the label if you are unsure.

Borax is a flux that has been used for many years with great success for brazing brass, bronze, steel, and malleable iron. One of the more popular mixtures of this flux consists of 75 percent borax and 25 percent boric acid, in paste form. This anti-borax flux as it is sometimes called is suitable for brazing those metals mentioned earlier in this paragraph. Other fluxes should be used for different metals, however, since there is no universal flux.

Because the fumes from most fluxes can be harmful to your health, it is important that there be good ventilation in your workshop area. Some fluxes, such as those containing sodium cyanide salts, are very dangerous. Avoid breathing these fumes or even letting the flux come in contact with your skin. Always check the container label for specific cautions.

Most fluxes are sold in powder form. This is quite convenient as the flux can then be applied to the metal in two different ways. The first way is to heat up the end of a bronze rod and dip it into the powdered flux. The hot rod will cause the flux to adhere to the rod, and the rod can then be used as it is, with a coating of flux (FIG. 8-4). Powdered fluxes can also be made into paste fluxes, usually by adding water until the right consistency is achieved (FIG. 8-5). Powdered fluxes can also be made into paste fluxes, usually by adding water until the right consistency is achieved (FIG. 8-5). Paste fluxes are commonly applied to the metal with a small brush. It is important that you read the container label because some fluxes require that you mix the flux paste with alcohol instead of water.

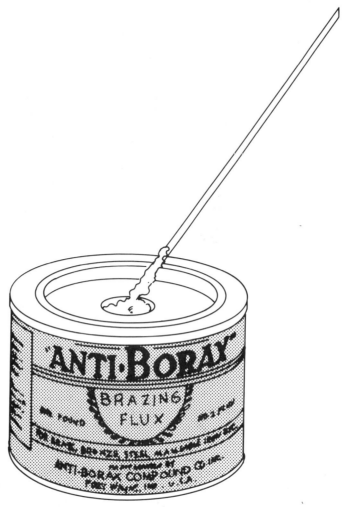

8-4 A hot brazing rod is dipped into the flux and is quickly coated.

PREPARING SURFACES

Easily, 75 percent of the success of any brazing project is directly related to the condition of the surfaces being joined. Not only must the surfaces be physically and chemically clean, but the pieces must fit together precisely. A joint clearance of 0.002 to 0.006 of an inch is acceptable for copper-alloy or silver-alloy brazing. For joining aluminum, greater tolerances are permissible: 0.006 to 0.015 of an inch. As you can imagine, joint design is quite important and will be discussed in this chapter. The clearances of a joint will have a direct bearing on the strength of the finished project. There must be a certain amount of clearance so the nonferrous metal filler can be drawn by capillary action into the joint. If the clearance is too great, however, the capillary attraction will be lessened. The end result will be a weak joint.

8-5 Mix up water and powdered flux to form a paste.

Cleaning metal

There are several ways of cleaning metal before the pieces are fitted together for brazing. These include brushing by wire, emery cloth crocus paper, or steel wool, and in special circumstances, washing the parts with an acid solution. If the metal has just been machined or milled, all that is usually required is a light touch with steel wool. But if the metal is old or weathered, more surface preparation will be required.

If rust is evident, for example, you should begin cleaning with a wire brush (FIG. 8-6). In most cases this tool will remove all scale and oxidized metal from the surface. Next, the metal should be sanded with emery cloth or corcus paper to make the surface smooth and shiny.

It is not recommended that a grinding wheel or other harsh abrasive be used on metal that will be brazed. These tools have a tendency to impart scratches or a rough texture to the surface. Knowing that close tolerances between the metal parts are important, this type of cleaning might cause tolerances that are unacceptable, at least from the standpoint of brazing. A grinding wheel is suitable, though, for surface preparation of metals that will be braze-welded (FIG. 8-7).

If much scale is in evidence on steel or iron and it cannot be easily removed by a wire brush, steel wool, or emery cloth, it might be necessary to soak the pieces in a special *pickling bath*. This is simply a mixture of water and sulfuric acid. After the pieces have soaked in the bath, they must be rinsed in warm water until all of the acid residue is flushed away.

Working with acid can be dangerous. Proceed with caution. Your local welding supply house should have the stock ingredients for a pick-

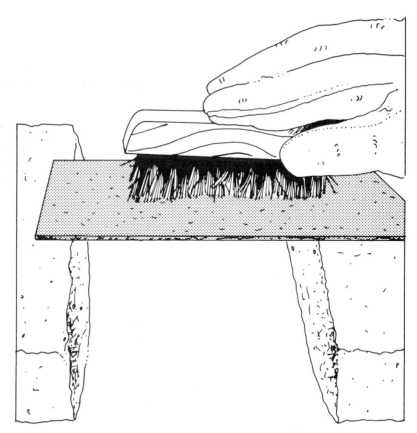

8-6 A wire brush is very handy for removing surface scale and dirt.

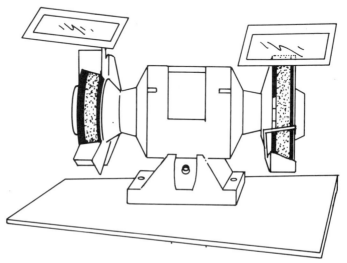

8-7 While a grinding wheel is not generally suitable for metal that will be brazed, it can be useful for metal that will be braze-welded.

ling bath and can probably offer some good suggestions as to how you might best go about this method of cleaning steel and iron. They might even perform this service for you. A pickling bath cleans metal like no other method. It is a useful means of cleaning not only small pieces, but also those metals that might require an enormous amount of work to clean.

Butt and lap joints

After the metal has been cleaned, the pieces must be fitted together so the tolerances are close: 0.002 to 0.006. The best way to make certain that the pieces fit together this closely is to use or develop a joint that is tight fitting. The two most common joints used are the *lap* and the *butt joint*. Whenever possible, the former is much preferred.

The main reason that the lap joint is popular is its strong joint. It is also relatively easy to make. Simply lap two surfaces together and braze. The length of the lap should be equal to at least three times the thickness of the thinnest member being joined. For example, if you are brazing two pieces of 3/8-inch-thick steel, using the lap joint, the length of the lap should be at least 1 1/8 inches. Figure 8-8 illustrates some of the more common joints, both good and bad.

Applying flux

After the pieces have been cleaned thoroughly, the joint surfaces must be coated with the proper flux. Remember that there are different fluxes.

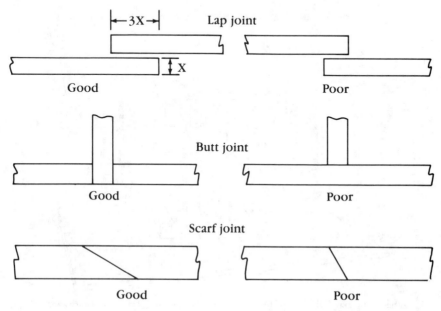

8-8 The three most commonly used joints in brazing.

You must choose the right one for the metal you are working with. In most cases, a pasty flux is brushed on the surfaces. If you are using a powder flux, follow the directions on the container (FIG. 8-9).

Once the pieces have been fluxed and fitted closely together, they must be held in this position until they are brazed and the filler material has cooled to a solid stage. A C-clamp can be used for this. As an alternative, a pair of locking jaw pliers can be quite handy. Locking jaw pliers are currently available in a wide variety of sizes and shapes. Your local welding supply house will probably have a selection of these to choose from as they are quite popular with professional welders. A standard 6- to 8-inch pair of locking pliers has hundreds of uses around the welding and home workshop.

The next step in the brazing process is, of course, joining the pieces with heat and filling the space between the pieces with molten nonferrous metal. Earlier in this chapter, I mentioned that careful surface preparation was about 75 percent of the task of brazing. The remaining 25 percent involves technique, which is covered next.

8-9 Paste flux is brushed on a joint that will be brazed.

BRAZING TECHNIQUE

Brazing is decidedly an oxyacetylene process, and the directions for this method of joining metal are discussed in oxyacetylene terms. As with all oxyacetylene metal joining, it is important when brazing to choose the correct tip size. The type and thickness of metal being worked on determines the correct tip size. Because some metals, such as aluminum, can be heated quite easily, a small tip should be used. It is important to keep in mind that the intention in brazing is not to melt the parent metal, but simply to get the joint edges hot enough to melt the filler material and make it flow. Also keep in mind that a neutral flame has about the same temperature with all size tips, but you might be able to have more control with a small tip. Probably the best place to look for the proper size tip for brazing is in the handbook that came with your torch. The manufacturer usually offers suggestions for tip sizes based on the types and thicknesses of various metals. See TABLE 8-1.

**Table 8-1 Welding tip sizes
and pressure settings for brazing.**

Metal Thickness (inches)	Tip Size (inches)	Size of Brazing Rod (inches)	Oxygen Pressure (psi)	Acetylene Pressure (psi)
1/32	1	1/16	5	5
3/64	2	1/16	5	5
1/16	3	1/16	5	5
3/32	4	3/32	5	5
1/8	5	3/32	5	5
3/16	6	3/32	6	6
1/4	7	1/8	7	7
5/16	8	5/32	8	8

Light the torch and develop a neutral flame that is appropriate for brazing. Some authorities suggest that a slightly oxidizing flame is better for brazing. In time you might feel the same. I suggest when you are relatively new to working with a torch that you work with a specific flame until you have mastered the basics of flame and torch control (FIG. 8-10).

Heating the base metal

The whole key to successful brazing is heating the base metal up to a temperature that will high enough to melt the nonferrous filler metal, but not cause the parent metal to become liquid. In most cases this is not as difficult as it might sound, since the melting point of all copper alloy brazing rods, brass and bronze, is between 1,100 and 1,800 degrees Fahrenheit. The melting point of most steels and iron is at least 2,100 degrees Fahrenheit.

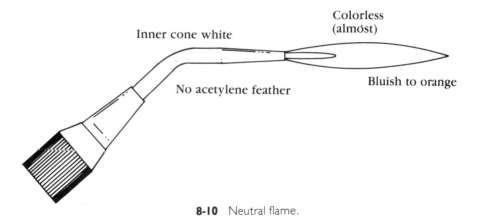

8-10 Neutral flame.

It is important that the joint be evenly heated. To accomplish this, it is almost always best to keep the torch moving in a circular, half-moon or zigzag motion over the area being heated (FIG. 8-11). If the joint is a long one, begin by heating up one section. After the initial section has been brazed, proceed along the joint, one section at a time.

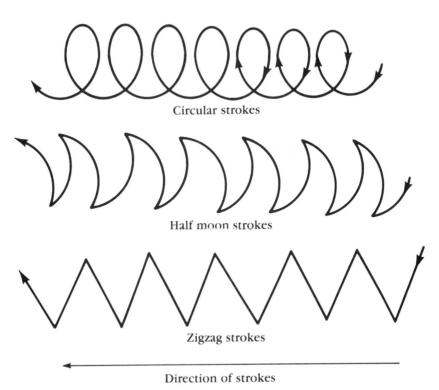

8-11 Common patterns of torch movement during brazing.

As you heat the metal, watch the flame closely through welding goggles, of course. Keep the cone of the neutral flame close to the edges of the joint but not touching the metal. The tip of the cone is the hottest part of the flame and can melt the metal if it comes too close to the joint. Overheating is a common problem for beginners (FIG. 8-12).

As the metal heats up, the flux around the joint becomes a good barometer of the approximate temperature. This is another good reason for choosing the proper flux for the metals being brazed. When heated properly, the flux will boil at about 212 degrees Fahrenheit, leaving a whitish powder on the metal surface. This powder will become puffy and begin to bubble at around 600 degrees Fahrenheit. It will become liquid on the surface at about 800 degrees Fahrenheit. At a temperature of about 1,100 degrees Fahrenheit, the flux will once again become liquid and darkish in color. This indicates that just a bit more heat is required to melt the filler rod. At this point, the metal of the joint will have a dull red color, which is another indication that the temperature is about right to melt and flow the filler rod.

Bonding

When the metal is approximately the right temperature, move the flame away from the joint and touch the end of the filler rod into the joint. If the metal is hot enough, the rod will flow over the heated area and into the joint (FIG. 8-13).

It is important to keep in mind that both pieces of metal must be of equal temperature or the filler rod will flow only to the hotter piece. This is the main reason for moving the flame over the joint in a circular (or other suitable) motion.

Once the filler metal flows over the heated area, bonding takes place. Don't disturb the brazed joint until the metal has cooled. This will ensure that the bond is a sound one.

After the joint has cooled, clean up the braze. The amount of cleaning necessary will depend on the parent metals, the filler nonferrous

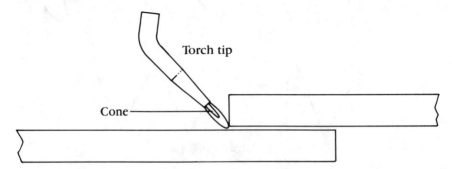

8-12 It is important to keep the cone of the neutral flame close to but not touching the parent metal. Keep the torch moving over the parts of the joint to distribute the heat evenly.

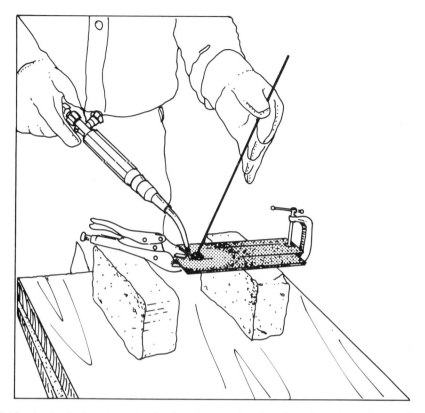

8-13   Apply the brazing rod only after the metal surfaces have been heated up to the point where they will melt the rod.

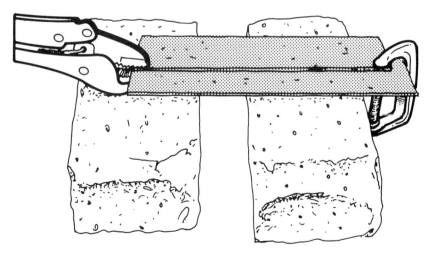

8-14   The work to be brazed must be held securely until after it has been brazed and the joint cooled.

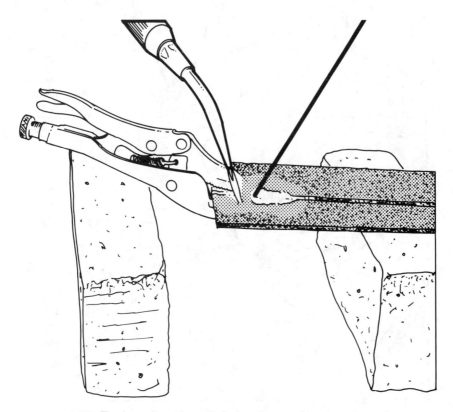

8-15   The bronze rod is applied after the metal has been heated.

metal, and the flux used. In most cases, only a slight slag or discoloration will be present on the brazed joint. Discoloration can easily be removed with special cleaners or a wire brush. If the slag buildup is heavy, you might have to resort to a chipping hammer and possibly a cold chisel (FIGS. 8-14 through 8-16).

BRAZING VERSUS FUSION WELDING

The obvious main advantage of brazing over fusion welding is that it is the best way to join different or dissimilar metals. Just about any two metals can be joined by brazing except aluminum and magnesium. Another strong point in favor of brazing is that, because the metals need not be brought to the melting temperature of the parent metals, there is almost never any distortion of the metals. This holds true when brazing is correctly done on any thickness of metal.

SILVER AND BRONZE BRAZING

In most instances, you will be brazing with a bronze filler rod. One alternative to this nonferrous metal is silver rod. Silver brazing differs only

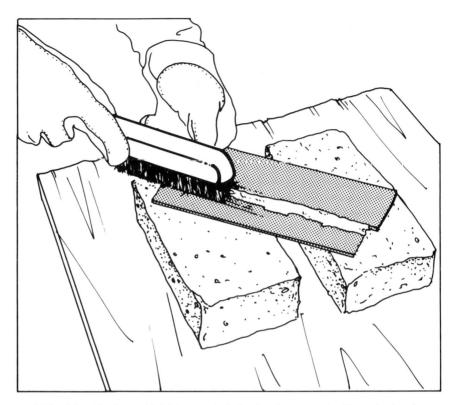

8-16 After the brazed joint has cooled, the flux is removed with a wire brush.

slightly from bronze brazing in that the melting temperature of the silver is lower than bronze. This means that less heat is required for the joining and the brazing can be accomplished a bit quicker. Silver rod, as you might expect, is more expensive than bronze. The technique for silver brazing is essentially the same as bronze brazing. A quick outline of these procedures might be handy for you at this time.

- Thoroughly clean the joint surfaces.
- Apply suitable flux to the joint area and only where you want the silver to flow. Silver rod will not adhere to any surface that is not flux-coated.
- Fit the pieces together and hold them in place.
- Heat the joint area evenly.
- Apply the appropriate brazing alloy (in this case silver rod).
- Let the braze cool undisturbed.
- Clean the surface.

ALUMINUM BRAZING

Aluminum can also be brazed in the same manner. Special aluminum brazing filler material and flux must be used for this, however, as brazing aluminum takes place at about the lowest temperatures of all brazing, 815 to 1,220 degrees Fahrenheit. Aluminum brazing requires the same surface preparation as other types of metals. Either mechanical or chemical cleaning should be completed before you attempt to braze aluminum. Many experts recommend chemical cleaning for aluminum because it is more thorough. Chemical cleaning consists of a series of steps that include dipping the pieces into a caustic bath, rinsing, dipping once again into an acid bath, and then rinsing. This chemical cleaning is repeated until the surface is totally clean. As with the acid pickling solution, the ingredients for chemically cleaning aluminum should be available from your local welders supply house.

Brazing aluminum is done at relatively low temperatures and a small tip size is often the best to use. This gives you more control, but not necessarily less heat. It is therefore very important to keep the flame constantly moving over the area being heated.

BRAZE-WELDING TECHNIQUE

The two major differences between brazing and braze-welding are the joints between the pieces and the fact that braze-welding does not use capillary action. Joint design for braze welding is similar to those joints used in fusion welding. Although the pieces must fit together snugly, the close tolerances necessary for successful brazing are not necessary for joining metals with the braze-welding process. In effect, the filler metal, which is nonferrous material as in brazing, is used to fill in the gaps between the pieces. When the filler metal comes in contact with the hot parent metal, it tries to open up the grain. As the filler seeps into the grain of the parent metal, a strong bond is achieved. To be sure, the pieces must fit together well but not necessarily to the close tolerances required in brazing.

Beveling metal edges

Braze-welding is a popular means of joining many types of metal by means of the butt joint. When joining metal that is thicker than 1/4 inch, it is necessary to bevel the edges of the joint (FIG. 8-17). For metal of lesser thicknesses, it is not necessary to bevel the edges. They should be chipped or ground slightly until bright metal shows. This will aid the filler in adhering well.

Probably the easiest and quickest means of beveling metal edges is on a grinding wheel (FIG. 8-18). You must work carefully, however, as grinding wheels can quickly remove metal. It helps to mark first mark the pieces and the angle of the bevel with a crayon or other suitable marking tool (FIG. 8-19).

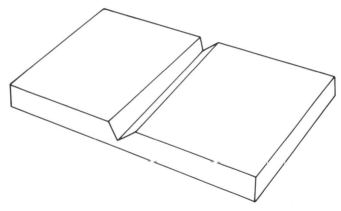

8-17 Metal ¹/₄ inch thick and thicker must have beveled joint edges for braze-welding.

8-18 Use a bench grinder to bevel the edges of metal that will be braze-welded.

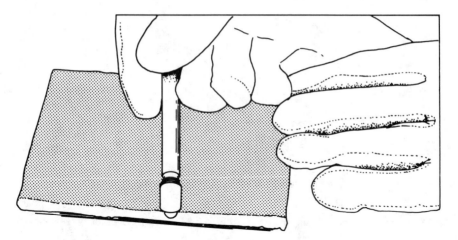

8-19 Mark metal with a soapstone pencil before grinding. Use the mark as a guide to the proper bevel angle.

Flame cleaning

All parts should be clean of all paint, grease, oil, or any other foreign matter. Solvents and a wire brushing will usually do the job. Another method used to clean metal of surface dirt is called *flame cleaning* (FIG. 8-20). This is simply a process where a neutral flame is passed quickly over the surface of the metal and burns off the dirt and other foreign material. This is not a recommended cleaning procedure for metal that has a coating of oil or grease; these will burn violently in the presence of pure oxygen.

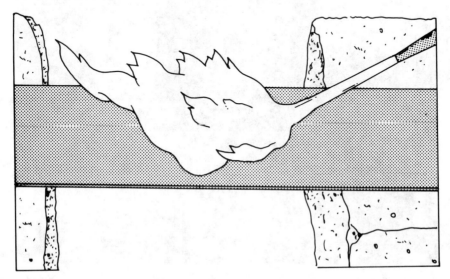

8-20 Flame cleaning metal.

After the pieces to be joined have been cleaned, they must be aligned and held in this position until the braze-welding is finished. This can be accomplished in the same manner as when holding pieces in a position prior to brazing, with clamps or locking pliers (FIG. 8-21). On large projects, the pieces are often tack-welded at this point, provided that this welding (in combination with brazing) is not undesirable (FIG. 8-22).

Preheating and heating

Preheating is often necessary in braze-welding to reduce the possibility of expansion and contraction stresses, which may be caused by the heat of the braze-welding operation. Preheating can be accomplished with a neutral flame. When the metal turns a dark red color, it is about right. Keep in mind that preheating is usually only necessary with the thicker pieces of metal, say over 1/4 inch.

8-21   Vise-Grips are handy for holding a joint steady while brazing.

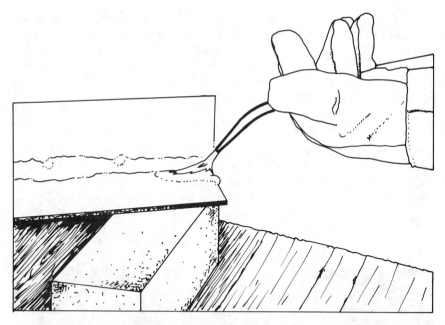

8-22 Tack-welding the edges of a joint.

It is important to keep the torch flame moving over the surface of the joints. This ensures that the metal is heated evenly along the joint. The flame, as mentioned earlier, is neutral, and the cone of this flame should not be allowed to touch the metal because the parent metal might melt.

While heating the joint surface, also heat the end of the bronze welding rod, and then dip it into the powdered flux. When the bronze rod is hot enough, the flux will adhere and coat the rod.

Tinning

When the parent metal is a very dark red, it is about time to push the end of the bronze rod into the joint. This allows some of the flux to melt onto the surfaces. As the parent metal melts the bronze rod, it forms a thin coat on the metal surfaces (FIG. 8-23). This is known as *tinning* and is similar to the tinning process in soldering. When tinning the surfaces of the joint in braze-welding, a rubbing action is the best way to get the flux and bronze to stick to the metal.

Be careful not to heat the parent metal too much because the filler metal will bubble, compromising good timing. On the other hand, if the metal is not heated enough, the filler metal will form little balls *on* the surface and not *in* the surface.

Once the surfaces of the joint have been completely tinned, you can go back over the joint and add more filler metal in one or more passes. It is very important that you work continuously, tinning and then building up the filler metal. If the filler metal is allowed to cool, it might break

8-23 A thin coat of bronze on a metal surface.

down when reheated. The end result will be a very poor quality braze weld (FIG. 8-24).

Cooling

Even cooling is important in braze-welding. There are several ways of accomplishing this, including heating the parent metal around the joint at least twice as wide as the distance of the joint. For example, if the width of the joint is 1/2 inch, heat the parent metal on both sides of the joint for

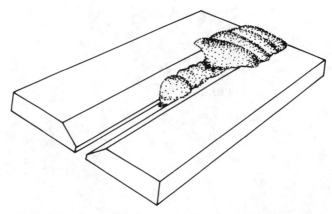

8-24 Joining thicker metal with two-pass brazing.

a distance of about 1 to 1½ inches away from the joint. Metal heated this way will cool slower than if just the joint area is heated.

Another way to slow down and make cooling more uniform is to cover the welded piece with asbestos cloth. This causes the piece to cool very slowly.

Still another way of cooling, and a method often used in industry when working with cast iron, is to pack the hot project in ashes or lime. This almost guarantees that the metal cools very slowly, thus creating more ductility in the welded material.

After the metal has cooled, it might be necessary to clean up the welded area. This is done to remove any flux residue or slag. A wire brushing is usually all that is required for this cleaning.

BRAZING TIPS

Brazing and braze-welding techniques are relatively easy to learn and master, providing you devote the necessary time to surface preparation and flame control. Undoubtedly, the best way to learn how to braze and braze-weld is to practice. Practice on scrap pieces of steel until you have a clear picture of what is necessary for success.

Brazing is, in effect, an extension of soft soldering, so this might be a good place to learn the basics. Begin by learning how to make simple lap joints using a soldering iron, flux, and solder. Work on sheet metal until you have an insight into the subtleties of heating metal. Once you have a handle on this method of metal joining, move on to thicker metal and use an oxyacetylene torch to braze the pieces together.

In the beginning, it might be easier to work on thinner gauges of steel, say ⅛-inch thickness or less. Again, practice the lap joint as it is probably the easiest to master.

- Clean the surfaces to be joined.
- Fit the pieces together closely and use the proper flux.

- Heat the metal evenly by constantly moving the neutral flame over the joint area.
- Keep an eye on the flux as this will be one of your best guides to the temperature of the metal.
- When the metal is hot enough, remove the flame and introduce the brazing rod. It should flow into the joint because of the temperature of the metal and not because of the heat of the flame.
- Let the brazed joint rest undisturbed until cool.
- Clean the cooled joint with a wire brush to remove flux and slag, if necessary.

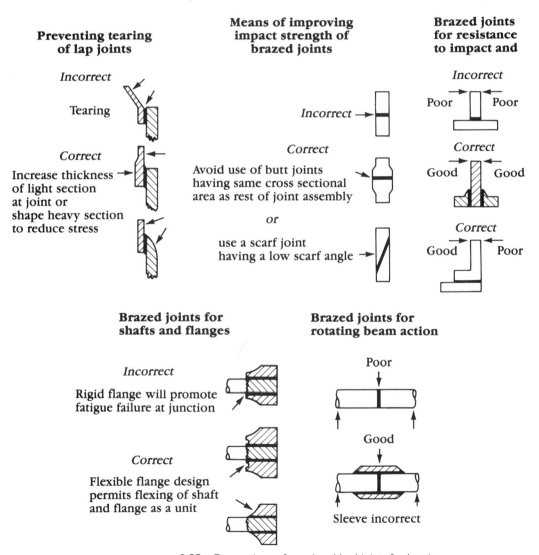

8-25 Comparison of good and bad joints for brazing.

After you have mastered the light touch necessary for successful brazing, and the lap, butt, and scarf joints, move on to braze welding. Remember that part of the key to braze-welding is applying a thin coat of the nonferrous filler material to the surfaces of the joint, known as tinning the surface. It is very important that this tinning coat be soundly attached to the surfaces of the joint metal. This requires bringing the surfaces up to the proper temperature before applying the filler rod.

In the beginning your brazed joints might not stand up under strain. As you become more familiar with the various metals and temperature ranges, however, you will be able to produce professional-looking finished joints (FIG. 8-25). Once you have mastered the techniques of brazing and braze-welding, the obvious next step is to learn how to weld.

Chapter **9**

Beginning welding

*U*p to this point, we have discussed two methods used for joining metal: soldering, which takes place at temperatures less than 800 degrees Fahrenheit; and brazing, which requires flame temperatures from 800 to just below the melting point of the metal being joined. In those methods, the metals being joined were never brought up to a temperature that exceeded their melting points. Now we will deal with the final method of joining metals, the one that produces the strongest attachment: welding.

FUSION WELDING

Four our purposes, the term welding is rather loosely used to describe the process whereby two pieces of similar metal are heated up to their particular melting point. The molten metal then flows together forming one homogeneous piece. In actuality, this process is called *fusion welding*.

A sound welded joint is a thorough mixture of the base metal and, in many cases, a filler rod made from a similar material. Regardless of the thickness of the metal, a good welded joint will be fused throughout the joint. When you are joining relatively thin pieces of metal, less than 1/4 inch thick, a thoroughly welded joint is not difficult. But to join thicker metals, special joint preparation techniques must be employed. As with both soldering and brazing, joint preparation is a necessary part of fusion welding.

Obviously, the best way to learn how to weld is to practice often and on various thicknesses and types of metals. It must be assumed, at this point, that you are familiar with your oxyacetylene equipment. You should know, for example, how to set up the equipment and develop the various types of oxyacetylene flames. These are the acetylene flame, carburizing flame, neutral flame, and oxidizing flame (FIG. 9-1). You should also have developed your own technique for keeping the torch flame in motion and at a specific height above the work. These techniques were discussed in chapter 8, but a quick review might be of help at this point.

	Neutral	Oxidizing	Carburizing	Aircowelding
Torch flames	Luminous cone 5850°F Envelope 3800°F 2300°F	6300°F	5700°F	5800°F
Ratio Oxygen acetylene	$\dfrac{1.04-1.14}{1}$	$\dfrac{1.15-1.70}{1}$	$\dfrac{0.85-0.95}{1}$	$\dfrac{0.92-0.98}{1}$
Effect on metal	Metal is clean and clear, flowing easily	Excessive foaming and sparking of metal	Metal boils and is not clear	Similar to neutral flame—little or no puddling necessary

9-1   Characteristics of oxyacetylene welding flames.

TORCH MOVEMENT

In order to ensure that the metal or joint between the two pieces of metal is heated evenly, the flame of the torch must be kept moving so that the metal will be brought up to a uniform temperature and will flow at the same time. This is best accomplished by moving the torch tip continuously. Figure 9-2 shows several methods or patterns, if you will, of moving the torch to distribute the heat evenly. At this time, at least one of these torch movement techniques should be familiar to you. More often than not, I find a circular movement the most effective and easiest to perform.

When welding metal, there is another technique that you will have to perfect. This technique also concerns torch movement, either forward or backwards, along the joint. Actually, there are two possibilities for general torch movement: the forehand welding and the backhand welding methods.

In the forehand welding method, the torch flame moves along the joint towards the tip of the welding rod. The position of the welding rod is actually on the cooler and not yet molten area of the joint. The flame of the torch is pointed in the direction of the welding, but also slightly downward so the flame preheats the joint area (FIG. 9-3). The forehand method works well with thin sheet metal.

The backhand welding method is the opposite of the forehand welding method. Actually the flame of the torch precedes the welding rod in the direction the weld is being made. For backhand welding, the welding

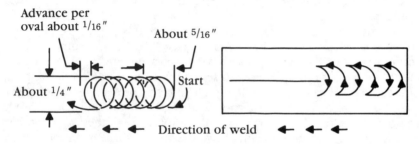

9-2 Two common blowpipe movement patterns.

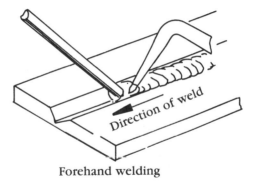

Forehand welding

9-3 Forehand welding technique.

rod is used to work the puddle of molten metal. As you can see in FIG. 9-4, this differs significantly from the forehand method of welding.

Generally speaking, the backhand method offers greater control of the welding process while at the same time providing welds with equal strength. In most cases it is the preferred technique for welding metal that is thicker than about 1/4 inch.

Throughout this chapter, unless otherwise stated, the flame being used for the welding process is a neutral flame. You might recall that the neutral flame is the third type, which is normally developed when a torch is fired up (FIG. 9-5).

As I mentioned earlier, the best way to learn how to weld well is to weld often. In the beginning of your welding, practice working on thin pieces of metal, about 1/8 inch thick. After you have mastered the various techniques necessary to weld metals of this thickness, move up to thicker pieces. Welding thicker metals is discussed in chapter 10.

One very good way to learn some of the basics of heating metal up to the melting point is to practice on a single piece of sheet steel. In the

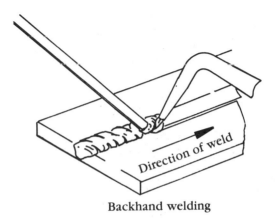

Backhand welding

9-4 Backhand welding technique.

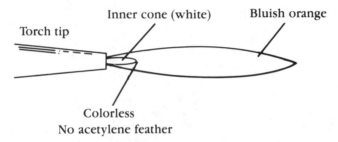

Torch tip
Inner cone (white)
Bluish orange
Colorless
No acetylene feather

9-5   Neutral flame diagram.

beginning, work with steel that is no more than $1/8$ inch thick. This metal is easily heated and is quite suitable for learning some of the techniques of welding.

PUDDLING METAL

For your first exercise, a 6- to 8-inch-long piece of steel across two fire-bricks. Set up your welding equipment and attach a small welding tip to the blowtorch handle. Consult the guide that came with your unit for the proper oxygen and acetylene settings for the tip being used. Light the torch and develop a neutral flame. It is assumed, at this point, that you have put on suitable clothing, gloves, and tinted welding goggles.

Heating steel

Begin heating the steel in about the middle of the piece (FIG. 9-6). You should hold the torch in a comfortable and balanced position. The tip

9-6   Begin heating the steel in the middle, and watch the metal change color as it gets hotter.

should be slanted at about a 45-degree angle to the right (southpaws should make suitable adjustments). The direction of travel will be in a circular motion from right to left. The circles you create should be about 1/4 inch in diameter. The tip of the cone of the flame should be 1/6 to 1/8 inch above the metal.

Watch the metal closely as it begins to heat up. It will turn cherry-red, orange, yellow, and finally white. About the time the metal turns white-hot, it is in a molten state. While heating the metal up to this state, it is important that you keep the flame cone moving over an area of about 1/4 inch. If you simply keep the flame in one spot, that spot will turn white and finally become a hole (FIG. 9-7). Since this is only practice, you should experiment with both holding the torch in a fixed position, which will cause a hole in the metal, and moving the torch in tiny circles causing the molten metal to puddle.

When the metal is in the molten state, you can move it around slowly with the cone of the flame. As you move the torch tip, you will notice that the molten puddle can also be moved. This, in fact, is one of the basic welding procedures. Much of your later welding will be carried out in a similar manner. In some cases, you will be using a filler rod, whereas in other instances you will simply be welding with the existing metal as we have just done in this puddling exercise.

Puddle characteristics

You can learn a number of things by carefully watching the puddle of molten metal. For example, the diameter of the puddle will be in proportion to its depth. On thin metal the puddle will be larger than on thicker

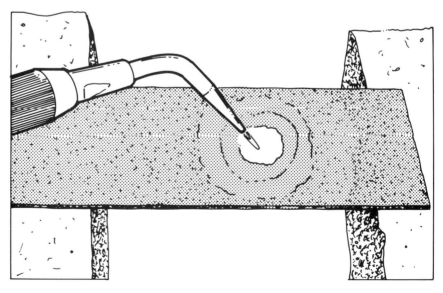

9-7 Hold the flame in one spot until you have burned a hole in the steel.

metal. In the latter case, you must keep in mind that the heat necessary to create molten metal has to penetrate through the thicker metal, which is a large part of the reason for the smaller puddle. Thicker metal also requires a larger tip size to help create more heat.

The general characteristics of the puddle will give you a very good indication of the type of flame you are working with. A neutral flame will produce a puddle that appears smooth and glossy (assuming, of course, that the grade of metal is a good one, with few impurities or dirt). If the puddle sparks or bubbles excessively, this is usually an indication that the metal is dirty or of poor quality. The puddle will appear dull and dirty if the flame is more carburizing than neutral.

When a properly adjusted neutral flame is being used to create a puddle of molten metal, you will notice a small bright incandescent spot on the edge of the puddle opposite the flame. This spot will move quite actively around the far edge of the puddle. If the flame is not neutral, the spot will be oversized and will barely move.

Once you have learned how to develop a puddle of molten metal and have burned a few holes in this scrap piece of metal, the next step is to move the puddle over the surface of the metal. A brief description of the techniques used will be of some help.

Welding beads

The torch tip, with a neutral flame, is $1/16$ to $1/8$ inch above the surface of the metal and held at about a 45-degree angle to the surface. The direction of travel will be from right to left and in a circular motion (FIG. 9-8).

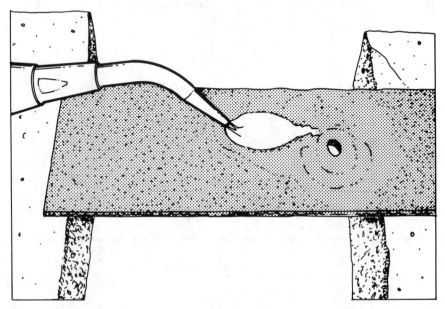

9-8 After you have burned a hole in the steel, try heating the metal once again in another spot. Try to puddle the metal in a straight line.

Begin forming a puddle just in from one edge of the sheet steel. As the metal begins to become molten, slowly advance towards the far end of the piece in a circular motion. As you move forward, the metal will go from a solid to a molten state. Then, as you move towards the opposite edge, the steel will become solid once again. You will notice that the circular movement of the torch will produce circular ridges as the metal cools. This line of ridges is commonly called a *welding bead*. Develop this welding bead across the face of the steel (FIG. 9-9).

With a little practice you should be able to create a welding bead that is uniform and straight. In the beginning, you will undoubtedly burn holes in the metal and probably develop a bead that more than slightly resembles a snake. After a bit of practice, however, you should be able to create a welding bead that is both uniform and straight.

Penetration or depth of a weld is very important. It is entirely possible to form a puddle of molten metal on the surface and not all the way through. When practicing, you should turn the scrap piece of steel over to see that the weld has penetrated through the metal.

As a side note, it makes a lot of sense to keep a pair of heavy-duty pliers handy while welding. The pliers can be used for picking up the hot metal. This will safely enable you to inspect the weld before the metal has cooled to the touch (FIGS 9-10 and 9-11).

You should practice puddling sheet steel until you can create a straight, well-penetrated, welded bead across the face of the steel. Most knowledgeable welders will probably agree that if you can create five straight and well-formed welding beads, one after another, you are probably proficient enough to try more serious welding (FIG. 9-12).

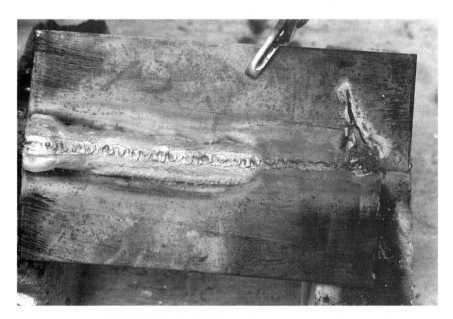

9-9 A finished weld done without a welding rod.

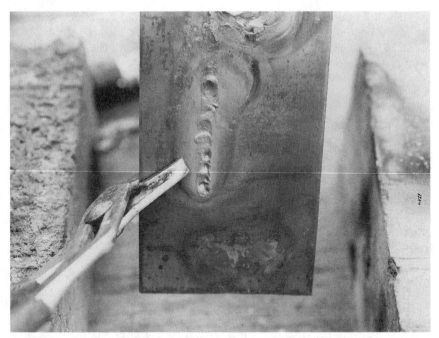

9-10 Pick up the steel with a pair of pliers and examine the cooled puddle.

9-11 Be sure to check the back side of the steel to see how well the puddle penetrated through.

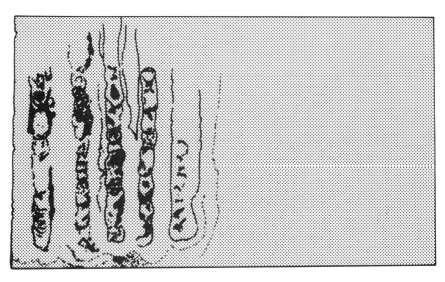

9-12 Practice puddling the steel with the torch until you can produce five straight and uniform puddles with good penetration.

WORKING WITH WELDING AND FILLER RODS

Until this point we have been practicing only with a single piece of steel and a flame. Since much welding uses a welding or filler rod as well, this will be our next objective. Since working with a welding torch and filler rod requires that you use both hands, the following exercise should not be attempted until you have reasonably mastered working with a torch to create a welding bead by puddling.

At this point, some mention should be made about welding or filler rods used by the oxyacetylene welder. Often during the oxyacetylene welding process, more metal than that which is melted by the torch is needed for a successful weld. To add more metal, it is common practice for the welder to use a welding rod that has the same properties as the base metal that is being joined, or very similar properties.

Welding rods are available in many different alloys and designed for just about any type of joining task. Probably the most common type of welding rod is simply mild steel. These are commonly sold in 36-inch lengths and are copper-coated to protect the rod from rust and corrosion while in storage. Uncoated mild-steel rods are also available. If there is an all-purpose welding rod, a mild-steel rod could rightly be given that name. Diameters range from $1/16$ to $1/4$ inch. Your local welding supply house will have a selection of welding rods. You should buy an assortment of these mild-steel, copper-coated rods. In most cases, you will use the $1/8$-inch-diameter rods with the greatest frequency (FIG. 9-13).

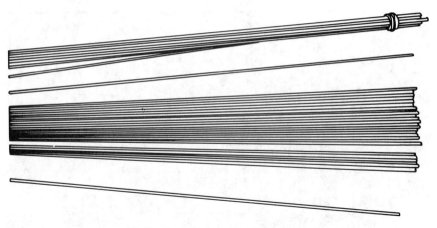

9-13   Assortment of welding rods.

Coat hangers

I should mention the economy of using a substitute to the all-purpose welding rod. If you spend enough time in the company of welders, you will probably hear that a standard wire coat hanger can be used as a substitute for mild-steel welding rod. While it might be true that a common coat hanger can be used as a filler material for welding, you should keep in mind that the steel used for these hangers is generally of a very low quality. It is therefore not wise to expect that a wire coat hanger will hold up under stress.

An acquaintance of mine was restoring an antique automobile. One of the many tasks that needed to be accomplished was the installation of new motor mounts in the engine compartment. It was about this time that the friend decided to purchase a new welding outfit and use it to help him in the process. After becoming reasonably competent at welding, he decided that it was time to start using the welding equipment for attaching new motor mounts. I must admit that he did an impressive job of welding the mounts onto the frame. The finished welds looked as if an old pro had done the work. The only problem with the project was that the welded mounts did not hold up under the torque of the engine because he used wire coat hangers instead of all-purpose welding rod as a filler material. To make a long and rather expensive story short, my friend had to reweld the mounts. The second time around, however, he used standard copper-coated steel welding rods and the welds have held to this day.

My friend is all that much wiser when it comes to welding. You can bet that he will never use wire coat hanger as a filler material again. In fact, in his entire household, you cannot find any type of clothes hanger other than a wooden one.

Adding filler material

To produce a welding bead with the aid of a welding or filler rod, you begin in very much the same manner as when working solely with a torch. The major difference, of course, is that in addition to holding the torch in your right hand, you will be holding (and moving) a welding rod in your left hand. The welding rod is used to add material to the puddle of molten metal. The more proficient you become at adding filler material to a weld, the better your overall welding and the stronger your welds will be.

It is important that you become proficient at adding filler rod to a puddle of molten metal. To practice, begin in the same manner as the first exercise in this chapter, with a single piece of sheet steel suspended between two firebricks. Hold the torch in your right hand and a welding rod in your left (FIG. 9-14). Begin heating the metal with small circular motions in about the middle of the piece of steel. As the metal begins to pass through the different color changes, slowly lower the tip of the welding rod into the flame to preheat it (FIG. 9-15). It is important that you do not touch the cone of the flame with the welding rod.

As the parent metal begins to become molten, the tip of the preheated welding rod is touched to the edge of the puddle. Almost immediately, the welded rod will melt into the puddle and mix with the parent metal (FIG. 9-16). After enough welding rod has been added to the puddle to create a slight crown on the surface, the tip of the welding rod should be withdrawn to about 1/8 inch above the surface of the welding bead.

Torch control is very important when you are introducing filler material to a weld. The flame should be in constant motion as should the tip of

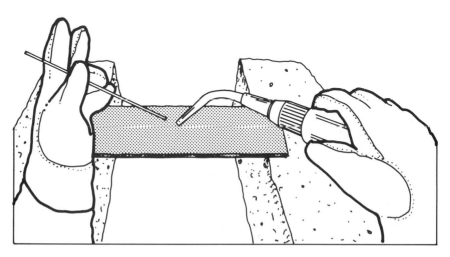

9-14   Practice holding the torch and welding rod in the proper position before lighting the torch.

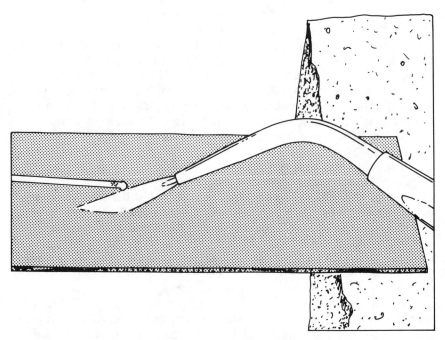

9-15 As you heat up the metal, keep the tip of the welding rod in the outer flame to protect it.

the welding rod. The cone of the flame does not cover a large area, but it is moving nevertheless. This will ensure even heat over a small area.

Practice melting filler rod into a weld bead until you can produce five straight welding beads. Probably the most difficult technique to master when introducing filler rod into a welding bead is the circular motion of

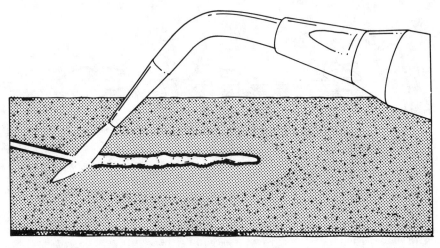

9-16 As the parent metal becomes molten, touch the preheated tip of the filler rod to the far edge of the puddle. The rod will then flow onto the surface of the metal.

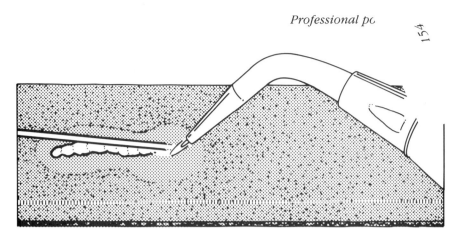

9-17 A good weld bead produced with a filler rod.

both hands. This technique can only be mastered through practice. It is extremely important that you practice adding filler material to the molten surface of the parent metal until you can produce a respectable welding bead (FIG. 9-17).

9-18 It is important to keep the tip of the filler rod inside the flame envelope but outside the cone.

PROFESSIONAL POINTERS

At this time a few professional pointers are in order. It is important that the tip of the welding rod be preheated before it is touched to the surface of the puddle of molten metal. A rod that is outside the flame envelope will be much cooler than permissible (FIG. 9-18). When it is touched to the puddle, it will cause the molten metal to cool. The end result will be a poor-quality weld bead.

On the other hand, if the tip of the filler rod is held too close to the cone of the flame, it will melt and drip into the puddle. More often than not, in such cases the filler material will be blown away from the weld, which will produce an uneven bead and, more importantly, incomplete fusion.

Torch control

Torch control when adding filler material is also quite important for a satisfactory weld. You should remember that the tip of the torch should be held at about a 45-degree angle to the surface of the parent metal (FIG. 9-19). When the torch is being held correctly with the proper flame, the surface of the puddle will be glossy and smooth. Often a slight angle adjustment is all that is required to produce a puddle with these characteristics.

Filler rod size

Another consideration for a successful and sound welding bead is the size of the filler rod. TABLE 9-1 offers some suggestions as to the diameter of the rod to use for the various thicknesses of metal. In the beginning, use the recommended welding rod for the particular job at hand. As you work with the different thicknesses of metal, you will develop a familiarity with how a particular diameter rod performs. You will therefore be able to create a suitable welding bead.

9-19 The torch tip should be held at a 45-degree angle, and the flame should be about 1/8 of an inch off the surface of the parent metal.

Table 9-1   Welding Rod Sizes

Metal Thickness (inches)	Welding Rod Diameter (inches)	Oxygen Pressure (psi)	Acetylene Pressure (psi)
1/16	1/16 – 3/32	4	4
1/8	3/32 – 1/8	5	5
1/4	5/32 – 3/16	8	8
3/8	3/16 – 1/4	9	9

Some points to consider concerning improper diameter welding rods are worth mentioning at this time. If a 3/32-inch welding rod is the proper one to use for a particular joint and you switch to a thinner rod, say 1/6 inch, you will find it very difficult to add enough filler material to the weld bead. A smaller rod will also make controlling the puddle and resultant bead very difficult. Moreover, the smaller rod will tend to oxidize or burn very quickly.

Using a rod that is larger than what is called for can have similar effects on the finished weld. For example, if a heavy rod is used on light metal, it might create too large a puddle and therefore a sloppy weld bead. A heavier than required filler rod might also tend to cool the puddle of molten parent metal too much while it is being added. The end result will be poor penetration and an ineffective weld bead.

JOINING TWO PIECES OF STEEL

Once you can create a straight weld bead on the surface of a single piece of sheet steel, it is time to begin practicing joining two pieces of steel. For this exercise select two pieces of 1/8-inch-thick steel strap, about 8 inches long. Lay the two pieces, with long edges aligned and butted together on top of two firebricks. In order to hold the two pieces together while welding, the ends should be tack-welded. *Tack-welding*, as the name implies, simply means to spot-weld one very small section of the pieces being joined. In addition to being a good way to learn how to heat up two pieces of metal to the melting point, tack-welding is a very useful means of holding two pieces of metal together while they are being welded along their length.

Tack-welding

Hold the lighted torch so that it balances in your right hand. Move the flame close to the end of the two strips, that area where the two strips butt together. Move the cone of the flame slightly in a tiny circular motion over the joint so as to cover an area of about 1/8 to 1/4 inch. Watch the metal change from its normal color to red and finally to white-hot.

Just about the time that the metal turns white in color, it should begin to flow together. As soon as this happens, you must move the flame away from the joint. It is important to keep in mind that if the heat continues to be applied to white-hot and flowing metal, you will often cause a hole to appear. This is undesirable. Your aim is simply to bring the edges of the two pieces of metal up to the melting point, and then just a bit more so the edges fuse together and become one. Circular torch movement usually helps in fusing the pieces (FIG. 9-20).

Once a tack weld has been completed on one end of the strips, move the flame along the joint between the pieces and back a few times to heat up the metal slightly. Then tack-weld the opposite end of the two pieces. While doing this tack welding, make sure that the pieces remain aligned.

In case you are wondering why the metal is heated up after one tack weld has been completed, this is so the two pieces will cool at approximately the same, even rate. This even cooling will reduce the chance that the strips will buckle or warp as they cool.

Prefitting pieces

There is nothing difficult about tack-welding. But there are a number of procedures used by professional welders to ensure the success of the weld. To increase your chance of producing a sound tack weld, you should know that the pieces to be tacked must fit tightly together along their entire length. Prefitting will make heating the two edges easier because you will not have to move the torch flame over a very large area. It will also enable you to flow the two metals a bit more easily. If the two edges are not aligned snugly, you can still produce a strong tack weld. But you will find it necessary to use a bit of filler rod to make up the difference between the pieces.

9-20   Tack-weld the joint edges of the two pieces without a filler rod.

Welding irregular edge pieces

In some cases, it might be necessary to make a tack weld with the addition of a filler rod. One example is when the two pieces being joined have irregular edges. To accomplish this, begin by aligning the edges of the two pieces as closely as possible. Working on one end of the piece, heat both edges up to the white-hot stage and add the filler rod. As you work, you will find it necessary to keep the flame moving constantly. Add the filler material at the exact moment the metal begins to flow. If you wait too long, the metal from either or both pieces might begin to drip, necessitating the addition of even more filler rod.

After the second tack weld has been completed, the strips will remain in position while you weld the long joint between the two pieces. Your first attempt at welding two pieces of metal together should be done without the aid of any filler rod. For the record, to tack-weld long pieces of metal (more than 12 inches in length), the edges are tack-welded along the entire length, about 6 inches apart.

Heat application

Begin at one end of the joint between the strips. With small circular movements of the torch, heat up the edges of the joint until they are in a molten state and are fusing together. As the joint begins to fuse, slowly move the flame toward the far end of the joint to ensure that complete fusion is taking place. It is important to keep the cone of the flame moving at all times, while at the same time working towards the far end of the joint. When working with thin metal, $1/8$ inch thick or less, you will find that you can move relatively quickly along the joint (FIG. 9-21). With

9-21 Puddle the edges of the joint without the aid of a filler rod. Keep the torch flame moving in tiny circles as the metal becomes molten.

heavier metal, you will find it necessary to work more slowly and with a larger tip.

In principle, this exercise in welding is the same as the first puddling-metal welding exercise in this chapter. The major difference, in this case, is that you are puddling the edges of two pieces of metal rather than simply working on a single piece of steel.

As you progress along the joint, it is important that you apply enough heat to fuse the edges of the pieces all the way through. Your best indication of the penetration of the weld will be the puddle being created by the cone of the flame. It will be helpful to you, especially at this early stage of metal joining, to stop welding and turn the pieces over to make certain that the weld has fused the edges of the pieces all the way through. Use a pair of pliers or a gloved hand for turning the metal over. If a circular weld appears on the back side of the joint, you are applying enough heat. However, if the weld does not show on the other side, you are not applying enough heat to the joint (FIG. 9-22).

WELDING JOINTS

Practice welding the joint between two pieces of light steel until you can accomplish this task with some competence. See if you can produce a

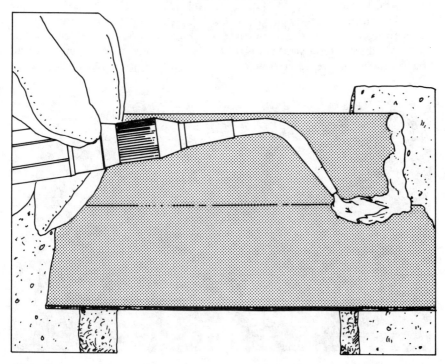

9-22 After you have made one puddling pass along the joint without a filler rod, turn the joined pieces over with a pair of pliers and check the penetration of the weld.

strong weld without the aid of a filler rod. After you have mastered welding two pieces of light steel, where the butting edges are on the same plane, you should turn your attention to welding other types of joints, still without the addition of a filler material.

Outside corner weld

Another type of welded joint that does not necessarily require the addition of a filler rod is called the *outside corner* weld. For this exercise, select two pieces of light steel, 1/8 inch thick, 8 inches long, and 2 to 4 inches wide. Place them together at right angles so that the vertical piece extends beyond the horizontal piece by about 1/16 inch (FIG. 9-23). Begin by tack-welding both ends of the strips to hold them in position while you weld the joint. The metal that extends will be used as a filler material as you weld.

A few professional pointers will help you to achieve success when making the outside corner weld. Begin by holding the torch so that the flame plays on the horizontal piece. This will cause the weld to form on the horizontal surface where it belongs. Cleaning up the weld bead will be much easier later on as you will only have to grind one side, the horizontal piece.

When making an outside corner joint weld, you will find that less torch movement is necessary than usual, particularly on butt joints (FIG. 9-24). You still must move the cone of the flame so that it heats up both pieces, but your main intention will be to "flow" the extended vertical piece onto the horizontal piece. This generally means less torch move-

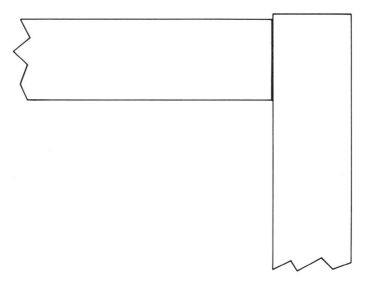

9-23 An outside corner weld with the vertical member extending 1/6 to 1/8 of an inch above the horizontal member.

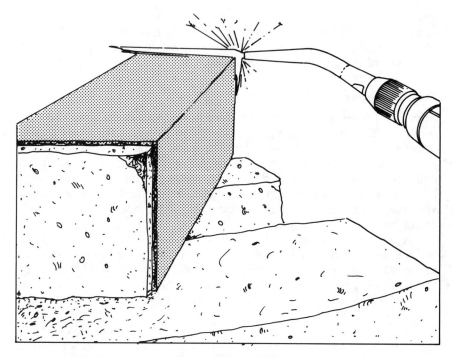

9-24 Proper torch position when welding an outside corner joint.

ment. You will also find that working from the outside of the vertical piece is the best way to approach this type of weld.

After the outside corner weld has been completed and the metal has cooled and solidified, check the weld first for appearance and then strength. A careful look over the weld bead will give you a good indication of how sound the weld is. Next, check the strength of the weld by trying to bend the two pieces of metal, much the same as you would open a book (FIG. 9-25). If you can hear any cracking or breaking of the metal, this indicates that the weld did not penetrate very deep into the joint between the pieces.

One of the unique features of the outside corner welded joint is that the weld need not be visible on the inside of the joints. In fact, the weld should be strong enough without penetrating all the way through. The best way to discover if the weld is, in fact, strong enough is to try to bend the two pieces apart. If the pieces hold without giving in the slightest, the outside weld is a good one.

Flange joint weld

Still another type of weld that can successfully be made without the aid of a filler rod is called a *flange joint weld*. To make this type of joint, you must first bend both edges of the pieces to be welded so that each piece will have an edge with a 90-degree angle. The flange should be from 1/4 to

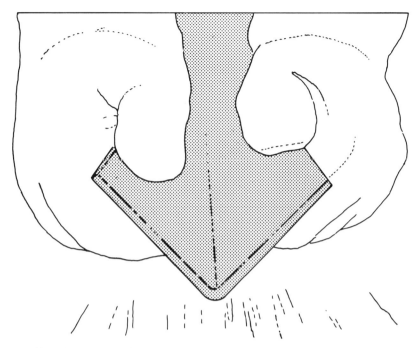

9-25 Test the strength of an outside corner joint by trying to bend the pieces, much the same as you would open a book.

$1/2$ inch wide and should be roughly equal to the piece it will butt against (FIG. 9-26). You will find a heavy-duty bench vise quite handy for bending the edges of sheet steel.

Begin a flange-welded joint by aligning the flanges of the pieces and tack-welding both ends to hold the pieces securely while welding. Next, heat from the torch is applied to the top of the flanges until they melt and fuse together. As the two flanges melt together, enough filler material from the parent metals should be present to produce a strong and sound weld (FIG. 9-27).

As you can see, there are a number of welds that can be accomplished without the addition of a filler material. Generally speaking, this is the case more often than not when working with thin steel. As the thickness of the steel increases, however, you will find it more difficult to develop a built-up welding bead. In fact, at this time you might have discovered that the bead you can produce by simply puddling the metal is slightly indented. Whenever this happens, it means that additional filler material is required. For our purposes, that will mean welding rod. It will probably be most helpful if I describe how to make some of the more simple welds using a filler material on relatively thin steel. Keep in mind that when adding filler material in the form of welding rod, we are approaching the joining process in very much the same manner as when joining metal without a filler rod, that is, puddling the metal. The major differ-

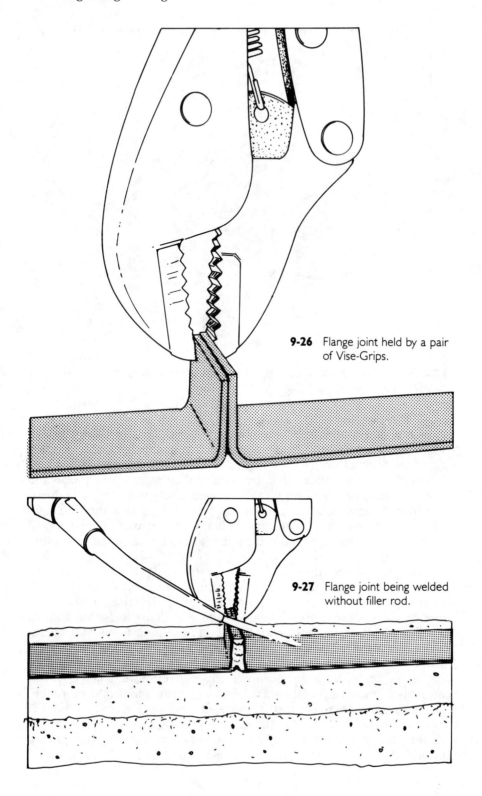

9-26 Flange joint held by a pair of Vise-Grips.

9-27 Flange joint being welded without filler rod.

ence, of course, is that now we are building up the weld bead with the filler material. Basically, everything else is the same.

Butt joint weld

Since the butt joint is the easiest of all joints to accomplish, you should begin welding with filler rod with this simple arrangement. Lay the two pieces of sheet steel, about 1/8 inch thick, side by side and suspended between two firebricks. Next, tack-weld both ends of the strips to hold them in position while you weld the joint. Begin, if you are right-handed, on the right end of the two pieces and work towards the left. For welding metal less than 1/4 inch thick, use the forehand welding method described earlier in this chapter. To refresh your memory, the forehand method of welding has the welding rod held in position over the as-yet-unwelded section of the joint. The flame of the torch is held at about a 45-degree angle to the surface of the joint. The flame is played on the joint in a circular motion, while the tip of the welding rod is held inside the flame envelope to preheat it. As soon as the metal edges of the joint become molten and begin to flow together, the tip of the welding rod is touched to the far edge of the puddle. This will cause some of the filler rod to flow into the joint and become part of the weld bead (FIG. 9-28).

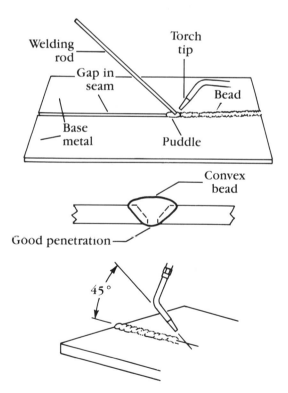

9-28 The filler rod is applied as the weld metal becomes molten and starts to flow.

You must move smoothly along the seam, constantly keeping the flame moving in order to heat both sides of the joint evenly. In the beginning you might find it difficult to hold the torch at a constant height above the joint, while at the same time holding the tip of the welding rod in the flame envelope. If you have a problem coordinating both hands, try locking your elbows against your sides and keep your forearms at a 90-degree angle to the front of your body. In time you will develop this technique. When you are first starting it seems difficult, hedging on the impossible (FIG. 9-29).

As you move along the joint, it is important to keep the flame moving at an equal distance over both edges. You will not be covering a very long

9-29 Try locking your elbows against your sides to steady both the rod and blowpipe actions.

span at any one time. Instead, make sure that you are creating a good pud-dle of molten metal. When the edges of the joint begin to fuse together, lower the tip of the preheated welding rod into the puddle. Allow the filler rod to flow into the puddle until the height of the welding bead is built up slightly. Then proceed a little bit further along the joint and repeat the heating. Your aim, when adding the welding rod, is to build up a joint that is equal along its entire length.

Whenever you come upon a tack weld, heat the area as if it did not exist, and add the filler rod. During the entire weld, the filler rod should be added in uniform amounts and at regular intervals. It is extremely important that you do not try to rush through the welding. Allow the edges of the metal to heat up sufficiently and to flow, so that good pene-tration and fusion can take place.

Practice the butt weld, with the addition of filler rod, until you can produce a respectable weld bead with good penetration through the joint. Work only with sheet steel that is less than 1/4 inch thick until you have perfected and mastered the torch and rod movements. After you can make a decent butt weld, the next exercise should be to learn how to make a T-*joint weld*.

T-joint weld

T-joints are formed by placing one sheet of steel on a similar piece of steel so that the two pieces resemble FIG. 9-30. The next step is to tack-weld the ends to hold the pieces in position while you weld the joints. After the tack welds have cooled for a few minutes, the next step is to weld the ver-tical piece to the horizontal piece. The forehand method of welding

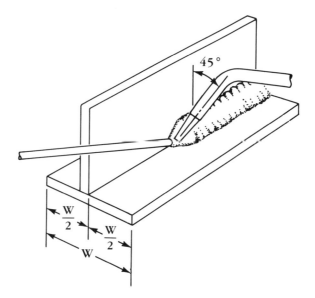

9-30 Welding a T-joint by using the forehand welding method.

should be used initially. Once you have reasonably mastered this technique, try the backhand method.

When welding a T-joint, it is important that you do not allow too much heat to build up on the vertical piece or you might burn through it. While the flame of the torch should be kept in motion during this welding, you will probably find that the pattern of movement will be very tight. Lower the tip of the welding rod into the joint when the metal of both pieces starts to flow. It is important at this time to keep an eye on the vertical piece as it will always have a tendency to reach a molten state quicker than the horizontal piece. If you find this to be a problem, adjust the angle of the flame so it is closer to vertical than horizontal (FIG. 9-31).

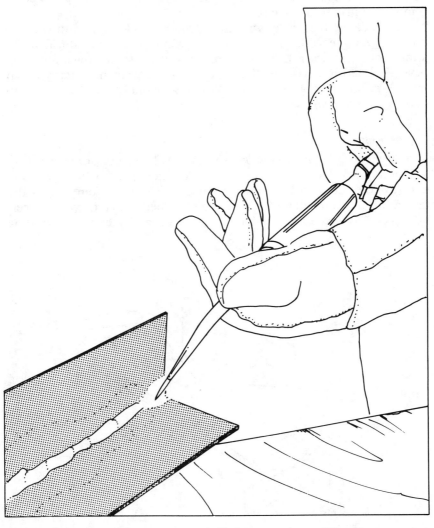

9-31 Adjust the torch angle more towards 90 degrees and you will have better control when making a T-joint.

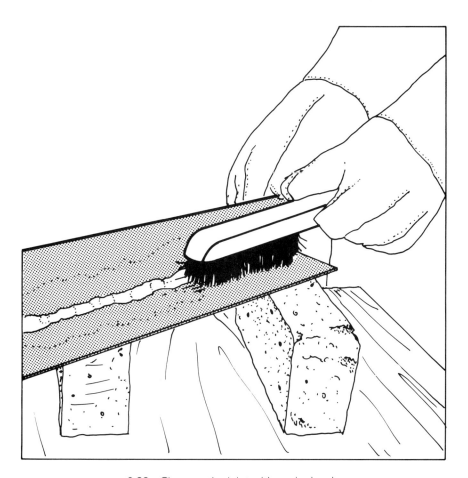

9-32 Clean up the joint with a wire brush.

After the joint has been welded uniformly along its length, allow the metal to cool and then inspect the weld. The bead should be built up to an equal height along its entire length and be equally distributed along both pieces. In other words, one side should not have more bead or filler material than the other. The vertical piece should appear to be solid and to have the same thickness as when you began. If the vertical piece looks thinner in certain areas, this is usually an indication that you have melted some of the metal away. The thinner area will not have as much strength as the rest of the joint (FIG. 9-32).

Chapter 10

Welding
thick metals

$\mathcal{C}$hapter 9 covered how to weld relatively thin pieces of steel, both with and without the addition of a filler material. In just about every instance, thinner metal does not require any edge preparation other than making sure that the joint fits together well, and that it is clean. When working with thicker metals, however, you generally cannot achieve good weld penetration unless you reduce the thickness of the metal along the joint area.

When welding any metal that exceeds 1/4 inch in thickness, you must reduce the thickness of the joint edges by grinding and then filling in the area with a welding rod. Actually, the point of edge treatment is to ensure that the heat from the oxyacetylene flame is even along the joint.

The thicker the metal along the edges, the more heat must be applied to make an edge metal that is molten throughout. As it turns out, metal thicker than about 1/4 inch cannot be heated thoroughly enough to make fusion welding absolutely complete. This is the basis for edge treatment.

There are four types of joints in all welding: butt, T-joint or fillet, flange, and lap. I have covered these in previous chapters, so I won't duplicate it here (FIG. 10-1).

BEVELING METAL EDGES

Thick metal requires that some type of edge treatment be initiated before welding begins. Often, this simply amounts to beveling one or more of the edges at about a 45-degree angle. Then, when the two pieces are placed together before welding, the beveled edges form a V, which can then be welded. In some cases, the V groove is deep and requires more than one pass with the torch and filler rod.

There are really only a few ways in which the edge of a piece of metal can be beveled. These include *grinding*, *cutting*, and *machining*. In most cases, only the first two are within the means and capabilities of the aver-

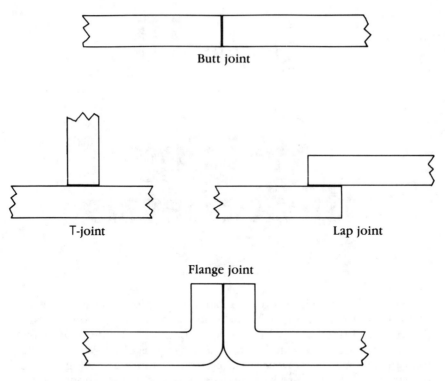

Butt joint

T-joint   Lap joint

Flange joint

10-1 Common joint designs for metals less than 1/4 inch thick.

age home welder, machining being a commercial production form of edge preparation.

Grinding

Grinding the edges of metal is probably the easiest and most efficient means of edge beveling. Because every home workshop should have some type of grinding wheel, chances are good that you already possess the equipment necessary for this special edge preparation.

First, mark the angle of the bevel, in most cases 45 degrees, and press the metal into the spinning grinding wheel. For larger pieces, you might find it easier if you set up some type of guide or rest on the front of the grinding wheel (FIG. 10-2).

One alternative to the conventional grinding wheel is a disc sander with a special, abrasive metal-cutting wheel or blade. This hand-held machine can be used for a variety of tasks around the shop. For example, it is handy for cleaning up metal before and after welding. A disc sander is also very useful for smoothing metal prior to finishing with primer and paint (FIG. 10-3).

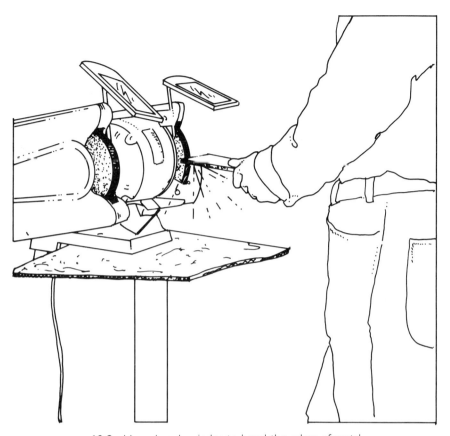

10-2 Use a bench grinder to bevel the edges of metal.

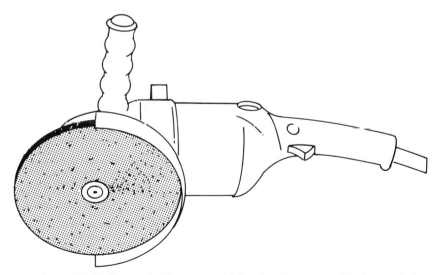

10-3 A hand-held disc sander is very useful for cleaning up metal before and after welding.

Cutting

Cutting with an oxyacetylene cutting torch is another way to bevel the edges of metal (FIG. 10-4). This beveling can often be done while the metal is being cut to size. It is important that the cut be of a high quality to minimize any additional metal edge preparation and to ensure that the pieces fit together well.

Unfortunately, cutting a beveled or chamfered edge on a piece of steel is not something that can be accomplished by a beginning welder. You will be better off to rely on a grinding wheel for your beveling work until you have become quite proficient at both welding and cutting with oxyacetylene.

WELDING STEEL PLATE

For your first exercise in welding thick metal, select two pieces of mild steel plate, 6 to 8 inches long and about 4 inches wide. Steel plate, in case you are wondering, is metal that is at least 1/4 inch thick. For this first exercise, you should choose plate that is about 1/4 inch thick. Steel of this thickness will be more than adequate for learning how to weld heavy metal.

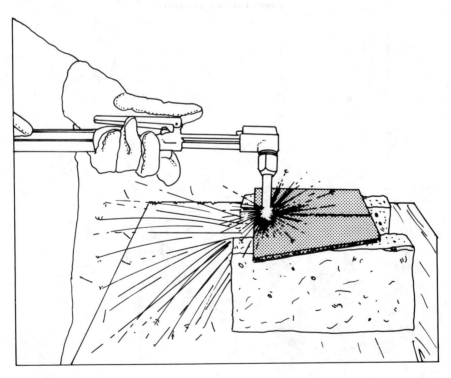

10-4   Beveled edges can also be made with a cutting torch. The whole key is to hold the torch at the right angle and steady.

Before you can begin welding, you must first bevel the edges that will butt together to form a joint. The best way to do this is on a bench grinder. Any quality bench grinder has a guide table in front of each wheel so that grinding can be done at an angle selected by the user. Grind one long edge of both pieces of steel plate to an angle of 45 degrees. It might be helpful at this time to explain how to grind the edge of a piece of steel plate.

Grinding plate edges

Make sure that the bench grinder is securely fastened to a work surface. Good lighting should be over the machine, and you should be wearing some type of clear eye protection. Lightly press one end of the plate into the spinning grinding wheel and move the piece with a back and forth motion. That way you will not grind one area more than another. Do not press the metal too hard into the wheel, as this will slow down the spinning and reduce the speed at which the metal is removed.

During the grinding, it is important that you keep the steel plate flat on the guide in front of the machine (FIG. 10-5). This ensures that the angle being ground is, in fact, consistent. Stop grinding often and carefully inspect the edge to make sure it is being ground at both the right angle and uniformly along the edge. Proper grinding takes time, so don't rush the work.

Tack-welding joint edges

After both edges have been ground, lay the two pieces of steel plate across two firebricks, with the beveled edges aligned. Tack-weld both ends of the joint to hold them in alignment while you weld the beveled joint (FIG. 10-6). Use a bit of filler material (welding rod) if you find it necessary for the tack weld.

As a side note, you should have already attached the proper-size welding tip into the blowpipe handle. When in doubt as to the proper-size welding tip, consult the handbook that came with your welding torch outfit. Here you should also find recommended working pressures for both oxygen and acetylene. See TABLE 10-1.

Begin welding the beveled joint at the left end, heating the edges of the joint thoroughly before introducing the welding rod. For a $1/4$-inch-thick steel plate, a $3/32$- or $1/8$-inch-thick, all-purpose welding rod should be adequate. It is important, when working with thicker metals, to keep the torch flame moving to produce consistent penetration. The size of the molten puddle will be the largest you have ever created, so give the metal time to heat up and flow properly.

Once the puddle is formed at the bottom of the V-groove, introduce the tip of the preheated welding rod. Flow the rod into the groove and build up the metal until it has filled the groove and risen to about $1/16$ inch above the joint. Continue along the joint until you have both filled the V-groove and built up the weld bead.

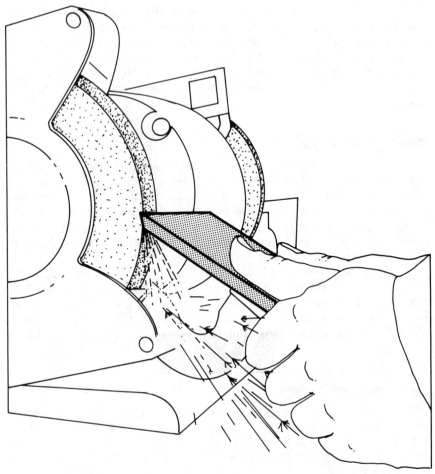

10-5 Use a guide to help you grind a true bevel on the steel plate.

Table 10-1 Pressure chart for Sears welding tips.

Metal Thickness (inches)	Tip Size (inches)	Welding Rod Diameter (inches)	Oxygen Pressure (psi)	Acetylene Pressure (psi)
1/32	1	1/16	5	5
3/64	2	1/16	5	5
1/16	3	1/16	5	5
3/32	4	3/32	5	5
1/8	5	3/32	5	5
3/16	6	3/32	6	6
1/4	7	1/8	7	7
5/16	8	5/32	8	8

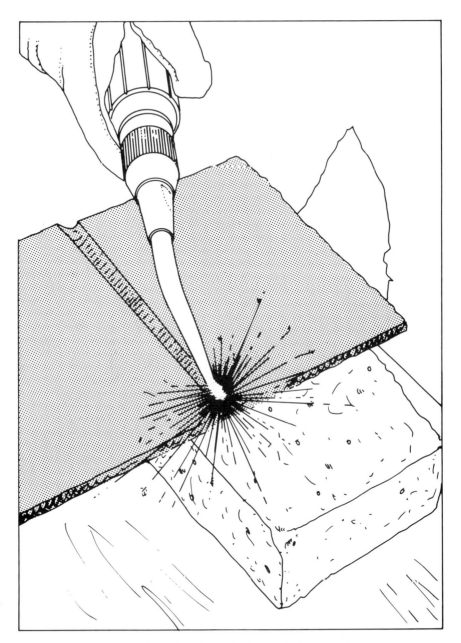

10-6 Tack-welding the ends of a beveled edged joint.

Welding steel plate is similar to welding sheet steel. The major difference is that you must work slower and add more filler material to produce a strong weld. By beveling the edges of the joint before welding, you are almost guaranteeing good penetration of the weld. One of the major points to keep in mind is that you must evenly heat up both sides of the joint to the molten stage before adding the filler rod material.

Strength test

After you have welded your first joint in steel plate, you should check the weld for strength. One very good way to do this is first to let the metal cool, and then to clamp the work in a vise and break the joint with repeated blows from a heavy hammer (FIG. 10-7). A sound weld will

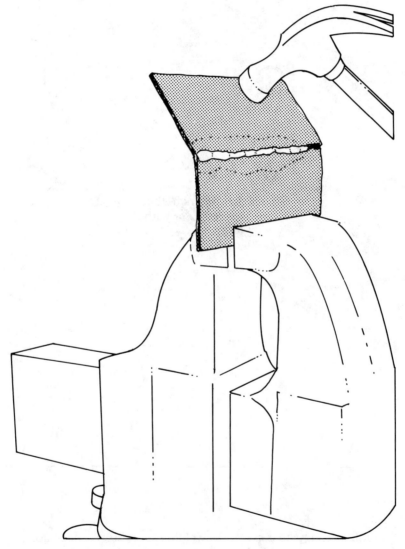

10-7 Test the strength of a joint by breaking it in a vise with blows from a hammer. A good weld will require a number of hammer blows while a weak joint will break quite easily.

require a lot of hammering to break while a poor weld will break quite easily. Practice the above welding exercise until you can produce a welded joint that is uniform and strong.

Characteristics of a good weld

It may be helpful at this time to describe what a good weld looks like when you are working with steel plate. The weld bead should be smooth and form an equal junction between the two pieces of steel plate. The width of the bead should be uniform and about 25 percent wider than the metal is thick. For example, if the steel plate is 1/4 inch thick, the width of the weld bead should be about 5/16 inch. The underside of the weld should be uniform as well, with no excess metal formations. In addition, the bead should be covered with a thin oxide film on both sides of the joint.

Figure 10-8 shows most of the joint possibilities for welding thick metal. Since these are time-proven joint designs, you should use them whenever possible. All these joints can be fashioned on a conventional home bench grinder, so the do-it-yourselfer should have little problem in duplicating them.

MULTILAYER WELDING

On very thick metal, greater than 5/8 inch, the home welder might find it almost impossible to fill the beveled joint with one pass of the torch. It is standard practice, in the welding industry, for the welder to build up the weld bead in several passes. Technically, this is called *multilayer welding*. By approaching heavy metal welding in this manner, the welder will find it much easier to deposit several layers of filler material until the welded joint has been built up to the proper height above the base or parent

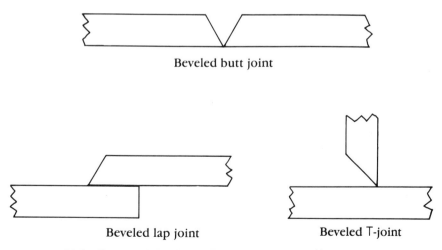

Beveled butt joint

Beveled lap joint   Beveled T-joint

10-8 Common joint designs for metal more than 1/4 inch thick.

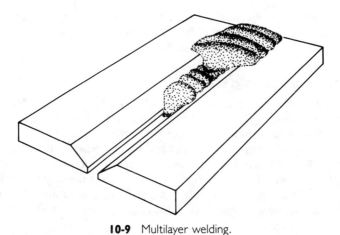

10-9 Multilayer welding.

metal. The point of multilayer welding is to produce smaller puddles of molten metal and enable the welder to have more control over the joint.

When building up a weld bead with multilayer welding, make sure the first layer provides good penetration at the bottom of the V-groove. Additional layers must fuse this first layer with the filler material and the sidewalls of the joint. The final layer seals the joint and should be crowned slightly above the base metal (FIG. 10-9).

Anyone knowledgeable in the art of welding will tell you that you should learn welding in gradual stages, mastering each before moving on. For the home welder, it would be best to begin with thin sheet steel. After mastering welding with and without filler rod, move on to steel plate. Do not attempt to weld plate thicker than 1/4 inch until you are able to produce a respectable welded joint. After you have the basics of torch control down, move on to thicker metal.

Chapter **11**

Finishing and problem solving

No welding project can be considered finished until the piece has been pressed into service. In some cases, this will mean simply cleaning up the weld to remove the ridges of the weld bead. In other cases, it might mean grinding the weld flush with surrounding areas and finishing with several coats of paint or clear finish. This chapter explains final protection techniques and examines some common welding problems.

PAINT COATINGS

The actual use of a welded piece will determine the extent of finish it receives. Keep in mind that ferrous metals (iron, steel, and alloys of these magnetic metals) will rust if subjected to the elements. Because rust will cause metal to deteriorate in time, you need to prevent this whenever possible.

A number of paint coatings are specifically developed for use over metal surfaces. You cannot cover metal with just any paint, such as latex wall paint, and expect this coating to perform a useful service. You should also keep in mind that the first coat is usually a special primer covering to provide a suitable adhesion base for the subsequent finish coats. Probably the best place to find more information about paint coatings is at your local paint store, or your welders supply store.

One alternative to a paint coating for welded metals is some type of oil finish. Simply stated, if you coat metals that will be exposed to the elements with oil or some type of silicone spray (such as WD-40), the coating will reduce the chances that the metal will rust. If you do this, keep in mind that the coating must be removed before the metal is ever heated again with a torch flame. Oil or grease burns violently in the presence of pure oxygen—a very dangerous condition.

SANDBLASTING

Before any coating is put on metal, the surface must be cleaned of any existing rust, flux, or scale. Some of the tools that will remove these surface coatings are a wire brush (hand-operated or mechanical), sandpaper, grinding wheel, or another suitable machine. Sandblasting is still another way to clean up a metal surface before finishing it. Sandblasting equipment is expensive, so it would probably be a good idea to have someone who specializes in this do the job. A good sandblast cleaning gives a truly professional look to your metalworking project.

COMMON WELDING DIFFICULTIES

Undoubtedly the most common problem encountered by beginners is popping the torch during the welding process. In most cases, this is a result of holding the cone of the flame too close to the molten metal. As you develop skill at keeping the torch tip at a consistent height above the work, you will find popping less of a problem.

Torch popping

Another cause of popping—and one that often makes the flame go out— is a welding tip that becomes dirty or clogged from bits of molten metal. The first thing you should do when the flame pops out is to turn off the acetylene and oxygen, in that order. Next, check the tip to see that it is not clogged. If it is, clear it by using a special tip-cleaning tool. Use a cleaning wire that is just smaller than the opening. Push the wire in and out of the tip hole several times. It is important that the wire be pushed in and pulled out straight in order not to enlarge or distort the hole in the tip (FIG. 11-1). Also remove any accumulation of metal slag around the orifice at the end of the tip.

If you move too quickly along a joint or if the oxygen/acetylene pressures are too low, the bead result be narrow and uneven. When inspecting your practice welds, and subsequent welds as well, keep an eye on the quality of the bead. A good weld well will resemble those shown in FIG. 11-2.

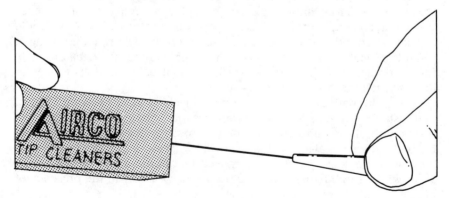

11-1   Clean your welding tips with a special tip-cleaning tool.

Range of thickness	Number of layers	Method of depositing	Cross section of weld
3/8″ to 5/8″	2	Two layers one pass	
5/8″ to 7/8″	3	Three layers two passes 1 & 2 in one pass layer 3 one pass	
7/8″ to 1 1/8″	4	Four layers three passes 1 & 2 in one pass layers 3 and 4 in two passes	

11-2 Suggested methods for multilayer welding.

Weld bead holes

The importance of constant torch movement cannot be overstressed. Holes in a weld bead are caused by holding the cone of the flame in one spot for too long (FIG. 11-3). If holes appear while the metal is still in the molten state, they can usually be filled with a bit of welding rod, simply by touching the rod to the area and letting molten material flow into the hole.

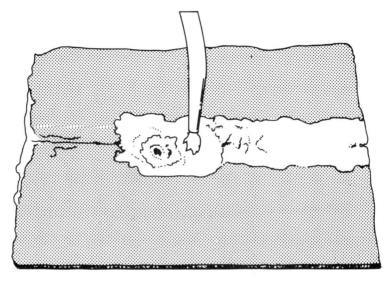

11-3 Holes in a weld bead are caused by not moving the blowpipe fast enough.

Strive for perfection in your welding exercises. A good weld bead should be a uniform series of circles with no ridges or high and low spots. This uniformity can only come as a result of a total familiarity with your torch, its adjustment, the filler material, and the metal being joined. Needless to say, it will take you many hours of working with a torch and the various types of metals before you can expect to produce good-looking as well as mechanically sound welded joints.

Warped metal

Metal has a tendency to warp when heat is applied. This is a problem that is more prevalent with light metal; in short, the thinner the metal, the greater the possibility of some type of warpage. Your awareness that metal will expand when heated and contract when cooled can be used to your advantage when you are working with relatively thin sheet metal. After welding, you can straighten the metal with your torch. If undesirable warpage presents itself, you can usually correct the problem by hammering and/or heating the opposite side of the warp (FIG. 11-4).

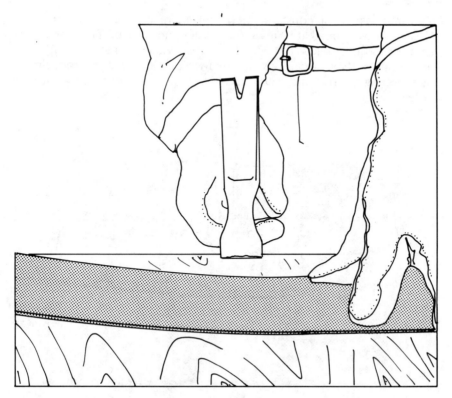

11-4 Warped metal can often be hammered back into the desired shape.

Chapter 12

The weldability of metals

*Y*ou already know from reading chapter 5 about the general characteristics and properties of numerous metals and their alloys. What you probably do not know is how easy or difficult it is to weld these different metals. The purpose of this chapter, then, is to supplement the information in chapter 5 and let you know what you are in for when you plan to weld a specific type of metal.

IRON

Wrought iron is very easy to work with because it approaches the melting point gradually and is easy to control. Wrought iron can be welded satisfactorily with all-purpose steel-welding rods. Also, low-carbon-content steel-welding rods are designed specifically for welding wrought iron. You should use this type whenever possible.

Cast iron is a chromium alloy and can be welded very much the same as mild steel, with a neutral flame. Actually, there are two types of cast iron: gray and white. The gray type of cast iron is the easier to work with. In about every instance, it is better to braze-weld cast iron. Square cast-iron filler rod, rather than round, is also available for welding cast iron.

STEEL

Steel, in one form or another, is probably the most common metal for the do-it-yourself welder. As you know, steel is classified according to its carbon content and/or its alloying element, such as chromium nickel, manganese, or molybdenum.

Mild steel or low-carbon steel is the easiest type of metal for the home welder to work with. Low-carbon steel is probably the best steel to use when working with the oxyacetylene process. Use mild-steel, all-purpose welding rods for best results.

High-carbon steel can be difficult to weld for the inexperienced hand because you must work quickly and keep the heat as low as practicable. All-purpose mild-steel welding rod can be used for joints in high-carbon steel. Stronger welds, however, can be achieved only by using special high-carbon filler rod.

Stainless steel is a difficult metal to weld for the oxyacetylene welder because it contains both nickel and chromium. The best method for welding stainless steel is TIG welding. If you find that you must weld stainless steel, keep in mind that it has a tendency to buckle and warp more than other types of steel. Reduce the chance that this will become a problem by tack-welding at close intervals along the joint before actually beginning the welding. You must weld stainless steel at an even, steady speed. One professional guideline to remember is that if the joint between the unwelded pieces begins to open as you weld, you are moving too slowly. On the other hand, if the joint begins to close in front of the flame, you are moving too quickly.

Low-alloy steels are generally quite easy to weld, especially when you are working with a sheet of alloy steel. These steels generally have a low (0.15 or less) carbon content. An all-purpose, mild-steel filler rod works the best.

Steel containing a large portion of an alloying element requires a special welding rod be used for a sound joint. Obviously, the filler rod should contain approximately the same proportion of the alloying element as the parent metal. Your local welding supply house is probably the best source of information about special welding rods.

COPPER

There are two types of copper: *electrolytic* and *deoxidized*. The former is difficult to weld because it contains a small amount of oxygen. Is is usually brazed with a silver rod rather than welded. Deoxidized copper is easier to weld, provided a special deoxidized copper-welding rod is used as a filler material.

Part of the difficulty in welding both types of copper is that the metal does not readily flow until it reaches the molten state. In other words, the metal remains solid until it reaches its melting point, which is about 1,980 degrees Fahrenheit. Once this temperature is reached, the copper will immediately become liquid.

Bronze, as you know, is an alloy of copper. This metal is very easy to weld and is quite popular for metal sculptures. For the welding of bronze, the obvious filler material to use is bronze brazing rod. You must also use a special flux when welding bronze. The flux forms a protective coating on both the metal and filler material, which prevents oxidation of the parent metal during bonding. The flux also dissolves the oxides that might form and further helps in the flowing of the filler material. Generally, you should use just a tiny bit more oxygen in the flame when welding bronze. This will mean that the flame will be slightly oxidizing. Because the amount of excess oxygen will be influenced by the particular bronze you are working with, a bit of experimentation is necessary in the welding of this metal.

ALUMINUM

Aluminum is the most common metal in the world. It is also very difficult to weld by use of the oxyacetylene process. Part of the reason aluminum is hard to weld is that it has a low melting point, about 1,220 degrees Fahrenheit. Unlike other metals, aluminum gives no warning as it approaches the melting-point temperature. A special welding process, MIG welding, was developed specifically for working with aluminum because of the problems of welding with the oxyacetylene process.

Nevertheless, aluminum can be welded with oxyacetylene, but it is a chancy proposition. Begin by preheating the metal with an excess acetylene flame. This flame will deposit carbon on the surface and make it look very black. Next, heat the joint area with a neutral flame until the carbon is removed. At this point, apply a special aluminum welding flux which is liquid. Then, while keeping the cone of the flame at least 1 inch above the joint, introduce a special aluminum welding rod into the flame. The object here is to tin the joint with the aluminum rod. After the joint has been tinned, melt more rod onto the area and build up the joint. It is important that the flame of the torch not play on the joint because it can make the aluminum weak. After the weld has been completed, all flux must be removed from the joint and surrounding area. This is usually done with warm water and a brush.

No special talent is required to learn how to weld. What is required, however, is a great deal of patience, along with a determination to master the basics of flame control. You should practice on lighter sheet steel in the beginning and not move on to thicker metals until you have mastered the basics of torch and flame control. In time, you will develop a certain sense about how metals are affected by intense heat, how metals flow, and just how much heat is necessary to manipulate molten metal. Practice until you know the characteristics of one metal; then expand your knowledge by experimenting with other types of metal.

Chapter **13**

Cutting metal

Undoubtedly, the most dramatic use of oxyacetylene equipment is in cutting metal. Sparks shower from the work and the cutting progresses very quickly. The equipment necessary for cutting metal with the oxyacetylene process is basically the same as for welding. The major difference is that the blowpipe must be fitted with a special cutting attachment. In addition, the working pressures are generally greater than for standard welding (FIG. 13-1).

EQUIPMENT

In most cases, a special cutting attachment is fastened onto the end of the standard welding blowpipe handle in place of the welding tip. It is to your advantage to understand the principles behind the operation of such attachments (FIG. 13-2).

Cutting attachments

All cutting attachments are basically the same, although design features often differ among manufacturers. A cutting attachment housing has three tubes running through it. One tube carries acetylene gas to the tip. Another similar-sized tube carries oxygen to the tip. A third, larger tube carries oxygen to the tip. The first two tubes carry oxygen or acetylene to a special expansion chamber just behind the cutting tip where the gases are allowed to expand and mix thoroughly. After this, they pass through the small holes in the cutting tip.

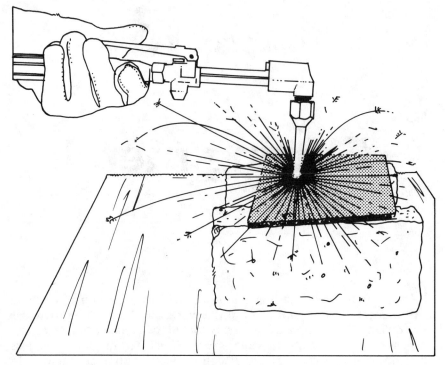

13-1   Cutting in progress.

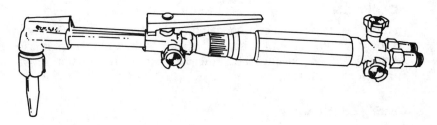

13-2   A typical welding torch with cutting attachment.

There are usually six holes in a cutting tip, which surround a larger hole in the center. The third tube in the cutting attachment handle carries pure oxygen, and it leads directly to the large center hole. The flow of pure oxygen through the tube to the center hole is controlled and regulated by a lever that sits on top of the cutting attachment handle. No oxygen can flow through this tube—or out of the large center hole—until the lever is depressed (FIGS. 13-3A and 13-3B).

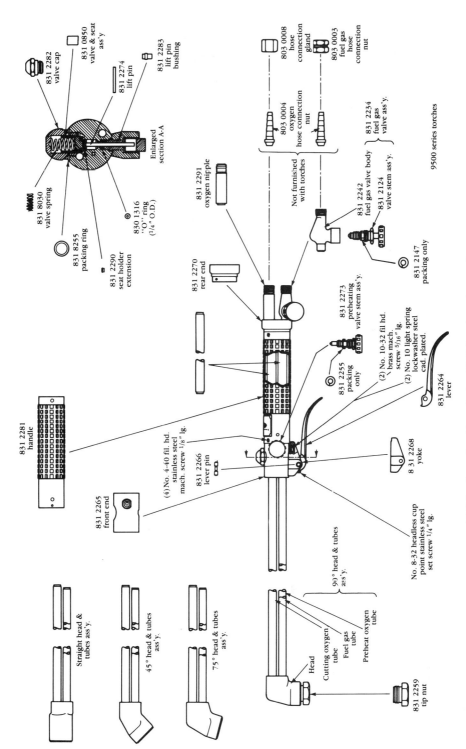

13-3A Parts for Airco Series 9500 hand-cutting torches.

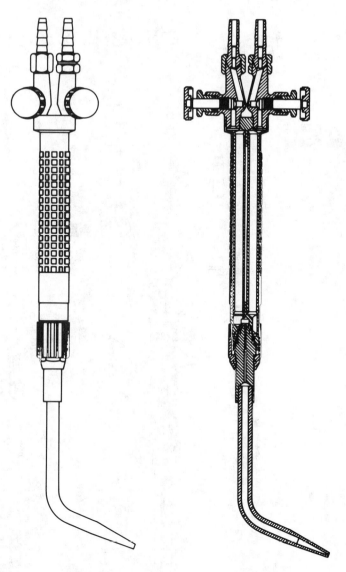

13-3B   Cross-section view of Airco Series 9500 hand-cutting torches.

Cutting tip holes

The six small holes in the cutting tip are called *preheat holes*, and the larger center hole is called the *cutting hole*. Each of the preheat holes produces a flame that resembles a standard welding flame. Since the oxyacetylene mixture flows equally through all of these small preheat holes, each flame will be the same: acetylene, carburizing, neutral, or oxidizing. As you can well imagine, the six preheat holes produce an enormous amount of heat (FIG. 13-4)

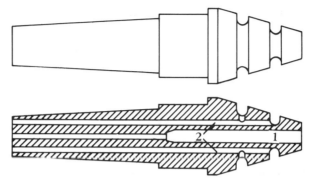

1. Cutting jet oxygen passage
2. Preheating-flame passages

13-4   Cross-section view of acetylene cutting tip.

The center hole in the cutting torch tip does not produce any flame. Instead, when the lever is pressed, a stream of pure oxygen is introduced into the heat zone created by the preheat flame. This causes the metal to oxidize very rapidly. It might be helpful at this time to explain how the oxyacetylene cutting process works.

CHEMICAL REACTION

Oxyacetylene flame cutting can be used for some ferrous metals. Oxyacetylene cutting can successfully be used to cut low-carbon steel, low-alloy steel, wrought iron, and other ferrous metals. This process does not work well for cutting stainless steel, cast iron, or any nonferrous metal such as aluminum.

The process of flame cutting is actually a chemical reaction. This reaction takes place because oxygen has a chemical attraction to ferrous metals when they are heated above their melting point. When excess pure oxygen is added to red-hot ferrous metal, the iron oxidizes very rapidly. To say it another way, the metal burns up. The force of the excess oxygen, in addition to burning up the metal, also aids in blowing away the molten metal.

SUCCESS FACTORS

Needless to say, there are several factors that influence the success of any oxyacetylene cutting process. It is important to use the proper-size cutting tip for the particular cutting project at hand. All manufacturers of oxyacetylene torches and cutting attachments offer a selection of various-size cutting tips, each designed for a specific thickness of metal. Obviously, you should use the recommended tip size for any given thickness of metal.

Each cutting tip will require specific working pressures for both oxygen and acetylene. In all cases the working pressure for oxygen will be con-

Table 13-1   Pressure chart for Sears cutting tips.

Plate Thickness (inches)	Tip Size (inches)	Oxygen Pressure (psi)	Acetylene Pressure (psi)
3/8 − 5/8	0	30−40	5−15
5/8 − 1	1	35−50	5−15
1−2	2	40−55	5−15
2−3	3	45−60	5−15
3−6	4	50−100	5−15

siderably greater than the working pressure for acetylene, as TABLE 13-1 illustrates.

The thickness of the metal being cut, in addition to dictating the cutting tip size, will determine the speed at which the metal can be cut. It is important to keep in mind that the metal is heated up to the "cherry-red" stage before additional oxygen is introduced. Obviously, thin metal will reach this stage much more rapidly than thicker metal. It is important, then, for an acceptable cut, that the welder keep a sharp eye on the area being preheated. You must watch for the metal to reach the proper temperature based on the color of the metal.

The preheat flames are kept operating during the entire cutting operation. The oxygen is introduced *only* when the metal has reached the proper temperature. If the excess oxygen is added before the metal has reached the proper temperature, oxidation or cutting will not take place.

As with welding, the best way to learn how to cut metal with oxyacetylene equipment is to practice often, and on various thicknesses of metal, until you have mastered some of the basics. It will be helpful, therefore, for us verbally to run through several exercises in flame cutting and point out the important functions that must be accomplished during the cutting operation. Before beginning, however, it might be helpful to quickly run through how to set up the oxyacetylene cutting equipment.

SETUP

Once your regulators have been attached to both the oxygen and acetylene cylinders, and the standard welding torch handle is also attached to hoses running from the regulators, remove the welding tip attachment. In most cases, this will entail unscrewing the welding tip holder from the end of the handle. At this time, the tanks should be turned off and the lines clear of both acetylene and oxygen.

The cutting attachment is then fastened to the end of the blowpipe handle and secured (FIG. 13-5). The next step is to fit the cutting attachment with the proper-size cutting tip for the project at hand. For your first exercise, you will be cutting light-gauge steel less than 1/4 inch thick.

Your welding outfit should have come with a booklet recommending the cutting-tip sizes for cutting various thicknesses of metals. Choose the

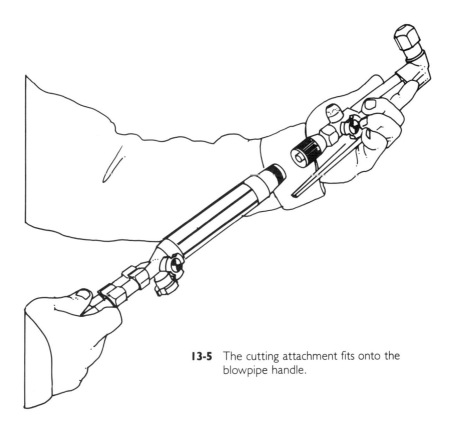

13-5 The cutting attachment fits onto the blowpipe handle.

right-size tip and fit it into the end of the cutting torch handle (FIG. 13-6). In all probability, you will be attaching a relatively small cutting tip.

Oxygen flow

Once the proper-size tip has been attached and you have double-checked all connections to make certain they are snug, open the oxygen tank valve slowly and allow some of the oxygen to pass into the regulator housing. At this time, the high-pressure gauge of the regulator should register how much pressure is in the oxygen tank. Open the valve slowly so that a sudden surge of pressure will not damage the internal parts of the regulator (FIG. 13-7). The diaphragm of the regulator should be wide open at this time. The low-pressure or working-pressure gauge should read zero. The control valves on the torch and cutting attachment should also be closed at this time. The next step is to set the working pressure for oxygen. This is done by simply turning the oxygen regular control lever until the low-pressure gauge registers. Set the working pressure according to the recommendations for the cutting tip being used. This information can be found in the manual that came with your welding outfit.

The next step is to give a quick check to all oxygen connections to make sure that no oxygen is leaking into the atmosphere. If all is well, open

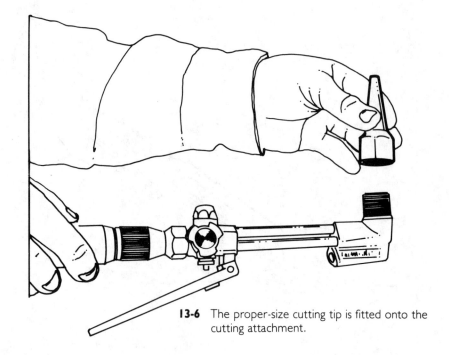

13-6 The proper-size cutting tip is fitted onto the cutting attachment.

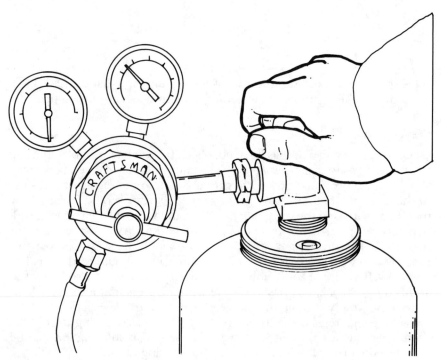

13-7 Open the valve slowly so the regulator is not damaged by a sudden surge of pressure.

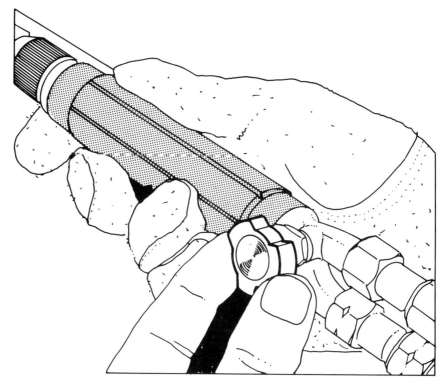

13-8 The oxygen control knob on the torch handle is opened all the way, but the flow of oxygen is controlled by the oxygen control knob on the cutting attachment.

the oxygen control valve on the welding torch handle two full turns. No oxygen should flow. The reason for this is that the oxygen flow is actually controlled by the oxygen control valve on the cutting torch handle rather than the valve on the welding torch itself (FIGS. 13-8 and 13-9).

Open the oxygen valve on the cutting torch handle a quarter turn to make sure the oxygen flows; then turn it off quickly. Next, depress the oxygen cutting lever for a second, and then release it quickly. When you

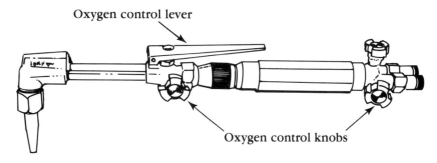

Oxygen control lever

Oxygen control knobs

13-9 The three controls for oxygen.

depress the oxygen cutting lever, a stream of oxygen should flow from the center hole in the cutting tip.

After the oxygen system has been pressurized and checked for leaks and proper operation, you can turn your attention to the acetylene line. First, make sure that the acetylene control knob on the torch handle is in the OFF position and the regulator control lever is also turned off (fully open). Open the acetylene tank valve slightly to let some of the fuel into the regulator housing. The high-pressure gauge should register how much pressure is in the acetylene tank (FIG. 13-10).

Acetylene flow

The next step is to turn the acetylene regulator lever until the working-pressure gauge registers the proper pressure. This information should be in the booklet supplied with your welding outfit. After the proper working pressure has been set, open the acetylene control knob on the torch handle to make sure acetylene is flowing (FIG. 13-11); then quickly turn it off. Give a quick check to the acetylene connections to make certain that there are no leaks.

As you can see, the only difference between setting up the oxygen and acetylene lines is that there are two control knobs for oxygen, one on the torch handle and another on the cutting attachment, but there is only

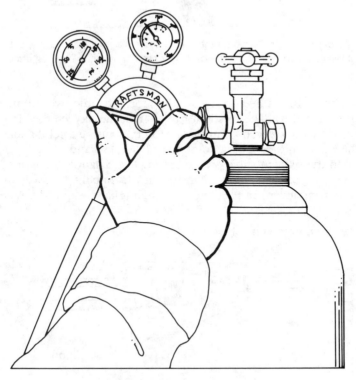

13-10 Set the working pressure for oxygen.

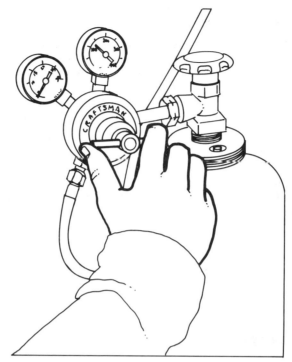

13-11 Set the working pressure for acetylene.

one for acetylene, on the welding torch handle. This might cause a bit of confusion initially, but I make it clear which oxygen control knob is to be used when cutting.

The flow of acetylene is controlled by one knob on the torch handle. The oxygen knob, next to this, is opened all the way during the cutting. It is important to remember that no oxygen will flow out of the torch cutting tip, however, until the oxygen-flow control knob valve on the cutting attachment is opened (FIG. 13-12).

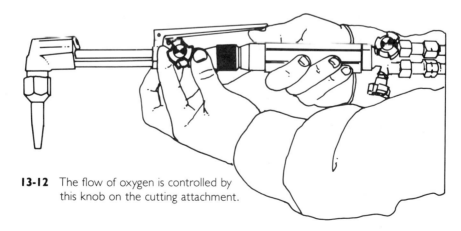

13-12 The flow of oxygen is controlled by this knob on the cutting attachment.

PRECUTTING PROCEDURES

The next step in learning some of the basics of oxyacetylene cutting is to lay your hands on a piece of scrap steel plate, about 1/4 inch thick, 4 inches wide, and 8 inches long. Lay this scrap lengthwise across two firebricks on your worktable. Now, with protective goggles in place, open the acetylene control valve a quarter turn, and light the torch. Open the oxygen valve on the cutting attachment. The valve on the torch handle next to the acetylene control knob is, of course, open already. Develop a neutral flame at the preheat holes, resulting in a cone about 1/8 inch long. You will find this easiest to accomplish by alternately increasing the flow of acetylene and oxygen until the proper flame is achieved. When you have adjusted it to a neutral flame, each of the preheat holes will have this flame (FIG. 13-13).

13-13   Adjusting to a neutral flame.

After the neutral flame is developed, depress the oxygen cutting lever momentarily to make certain that oxygen is, in fact, flowing through this part of the cutting attachment. You might find when you depress the cutting lever that the preheat flame needs to be adjusted. This is often the result of excess oxygen in the center of the preheat flames. Adjust the preheat flames if necessary. Now the torch is ready for the cutting operation.

Position the hoses

Before actually beginning to cut, take a quick look around the work area to make sure the oxygen and acetylene hoses do not run under the work. This is dangerous, especially if a spark or hot slag were to fall onto the hoses. It might cause the hose to rupture and a fire could result. For the record, it is always best for the hose to run from behind the welder. Your body will then always shield the hoses from hot sparks and molten slag (FIG. 13-14).

Check that you have a clear path to the oxygen and acetylene tanks. If a fire were to break out, turn off the flow of oxygen and acetylene at the tanks after turning off the control valves on the torch handle and cutting attachment.

Access to safety equipment

Safety equipment should always be within easy reach. This includes a fire extinguisher. See chapter 14 for information on fire-fighting equipment. A bucket of sand and bucket of water should also be in the welding shop for a quick response to any type of fire (FIG. 13-15).

13-14 Oxygen and acetylene hoses should come from behind the cutter.

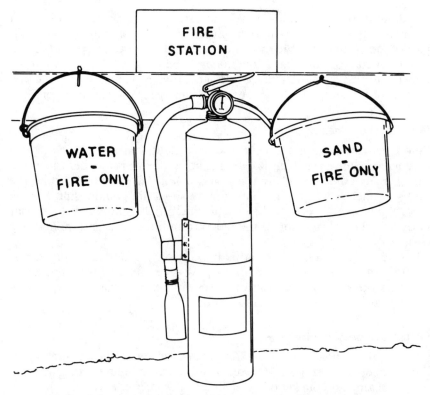

13-15   A fire station should be within easy reach of the cutting area.

You should have tinted eye goggles. Many experts suggest that the molten metal being cut is brighter than when welding. Make sure your goggles are dark enough to protect your eyes while at the same time allowing a clear view of the work in progress (FIG. 13-16). Heavy leather gloves and other body protection should also be worn because sparks and hot slag are a common occurrence in an oxyacetylene metal-cutting unit.

It is also important to wear the proper clothing when cutting or welding. Chapter 14 contains extensive information about the types of clothing that are suitable to wear when working in the welding shop.

CUTTING A HOLE IN STEEL PLATE

To get a feel for how a cutting torch can pierce through metal, the first exercise will involve cutting a single hole in the center of a steel plate. Begin by adjusting the torch for a neutral flame. Then add a bit more oxygen to develop a slightly oxidizing flame—which is better for cutting than a neutral flame. Hold the tip of the torch about $1/8$ to $1/4$ inch above the center of the steel plate. The idea is to hold the white cones in each of the preheat flames just above the surface of the metal. In a matter of

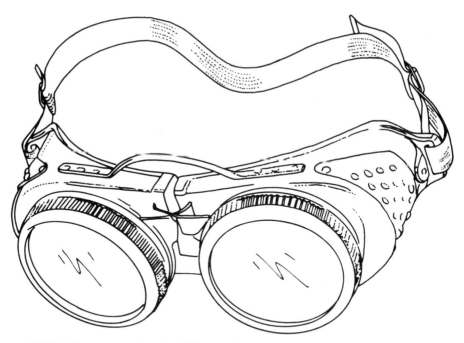

13-16 Always wear tinted welding goggles when working with a cutting or welding torch.

moments, you will notice the change in the color of the metal until finally the color is a "cherry-red." At this time, depress the oxygen cutting lever on the cutting attachment. It is important to press the lever slowly to blow a stream of pure oxygen onto the molten steel.

As the oxygen hits the molten metal, a shower of sparks will appear at the center of the cut. If you hold the torch steady, the oxygen will remove the surface layers of the metal very quickly, exposing the lower layers. These layers will almost immediately be heated by the preheat flames, and pre-oxygen will blow them away as well. In a very short period, you will blow a hole in the center of the metal. This is basically how all oxyacetylene cutting is done. The difference between blowing a single hole through steel and cutting a straight line across a two-foot piece of steel is simply a matter of practice and finesse (FIG. 13-17).

Some general comments about cutting are in order at this time. It is important that the cutting torch be held at a constant and steady height above the metal surface. It is helpful, especially in the beginning, to use two hands to hold the torch (FIG. 13-18). Assuming you are right-handed, your right hand will hold the cutting torch at the balance point and your right thumb will be on top of the oxygen cutting lever. Your left hand will be in front of your right and will serve as a steadying point for the cutting torch. Of course, both hands should be encased in heavy leather gloves to protect them from hot sparks and slag.

13-17 Just before the white-hot stage, press the oxygen cutting lever.

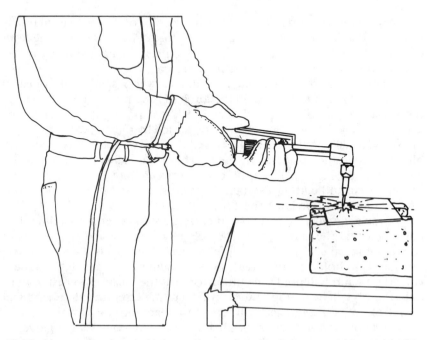

13-18 Use two hands to hold the cutting torch and to help you cut in a straight line.

One last reason for holding the cutting torch $1/8$ to $1/4$ inch off the surface is that, at this height, the preheat holes and oxygen cutting hole will be somewhat protected from reflected heat. At this height, the holes are less likely to become clogged with flying bits of molten metal.

CUTTING A LINE ACROSS STEEL PLATE

The next exercise involves cutting a straight line across the steel plate. If you did not burn too many holes in the piece of steel you used for the first exercise, you can use it for this exercise as well. Before cutting, mark the steel plate so you will have a guide for the cutting. The accepted method of marking metal in the welding trade so to use a special *soapstone* pencil (FIG. 13-19).

Soapstone is simply a type of talc that remains visible under heat. It is used by welders for marking straight and curved cutlines. Special pencils are available that can be refilled with lengths of new soapstone, and they work very much the same as a common mechanical lead pencil.

One alternative to using a soapstone pencil to mark cuts on metal is to mark the cutline with a series of indentations spaced about $1/4$ inch apart. These are made with a center punch and a hammer. While this alternative method works well, it will take you quite a bit longer to mark your cuts than if you simply use a soapstone pencil (FIG. 13-20).

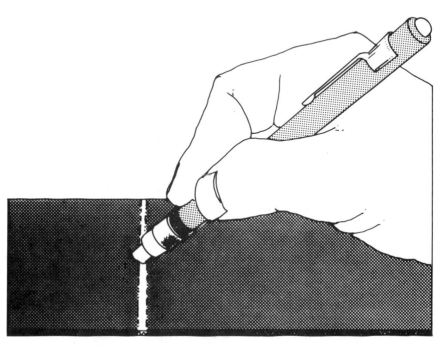

13-19 A soapstone pencil is the best tool for marking metal to be cut.

13-20   A punch and a hammer can also be used to mark a cutline on metal, but this will take longer than using a soapstone pencil.

Mark the scrap steel plate with a line running across the piece (FIG. 13-21). Next, position the steel plate so the cutline is about midway between the two firebricks holding up the steel plate. Then light the torch, develop a neutral flame, and begin heating the metal on one edge of the cutline.

Depressing the oxygen lever

Once the edge of the steel plate is cherry-red in color, the oxygen-cutting lever is slowly depressed. This allows a stream of pure oxygen to blast against the molten metal, resulting in sparks and the beginning of a gouge or cut.

The reason for depressing the oxygen lever slowly is so that the heated spot on the metal won't be cooled by the flow of oxygen. This cooling would prevent cutting from taking place (FIG. 13-22).

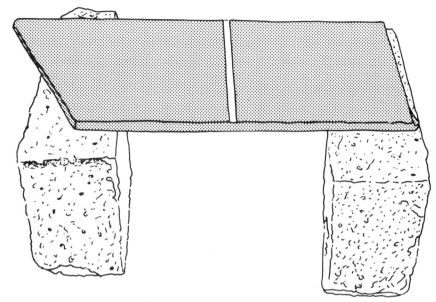

13-21 The cutline on a piece of steel plate.

After the metal has been gouged by a fine stream of oxygen, you will notice that the hole becomes deeper. It is at this time that you should press the oxygen lever once again to complete the cut. At about the same time, you should begin to move the cutting torch forward along the cut-

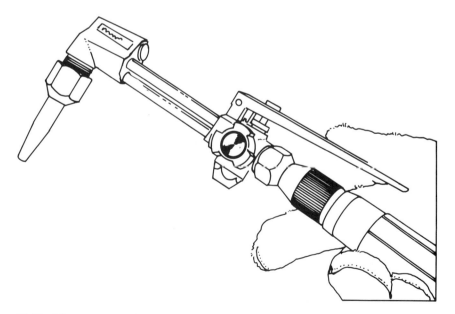

13-22 The oxygen cutting lever should always be pressed slowly and then held when the proper cutting action is reached.

line about ¹/₂ inch. This metal is very close to the metal that has just been severed with oxygen, and it will be very hot. You will only have to heat it with the preheat flames momentarily. Next, spray a little of the pure oxygen onto the area to cut the metal. As this happens, move the torch flames forward once again. Basically, this is how all metal is cut with an oxyacetylene flame and a stream of pure oxygen.

Cutting speed

The speed at which you progress along the cutline will be determined by a number of factors, the thickness of the metal being the most crucial. Use an even motion when cutting metal. It will take time to learn just how much heat is needed before pressing the oxygen cutting lever.

Continue cutting until you have made one pass over the steel plate, following the soapstone cutline. Turn off your cutting torch, and let the metal cool until it can be inspected. If you are in a hurry, you can drop the pieces into a bucket of water to rapidly cool them (FIG. 13-23). If you do this, it is a good idea to pick the hot metal up with a pair of pliers. I always keep a pair on the welding table for just such tasks. When I am working off the table, I often stick a pair of pliers in my back pocket for instant use.

13-23   Metal can be quickly cooled in a bucket of water.

After the metal has cooled, look closely at your first cut. There are, basically, four possibilities of how the cut will appear: the cut was either perfect, too slow, too fast, or involved too much heat. Each of these cuts has definite characteristics.

Perfect cut

A perfect cut will, at first glance, appear as if the metal has been cut with a hacksaw. There will be no, or very little, buildup of metal on either the top or bottom edge of the cut. The face of the cut will appear to have fairly straight lines running through the metal and each will be about the same depth as surrounding marks. In short, a good cut made with an oxyacetylene cutting torch will look good, almost as if it were made with a machine (FIG. 13-24).

"Too fast" cut

A cut through steel plate that has been made too quickly will have a buildup of slaglike metal on the bottom side of the plate. This is a result of not allowing enough time for the pure oxygen to blow the slag out and away from the metal. The cutlines on a piece of metal that has this slag buildup will be curved, almost as if the metal were cut with a circular saw (FIG. 13-25).

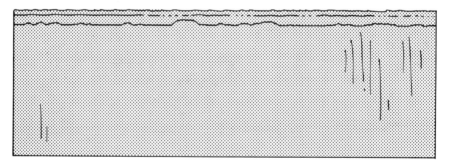

13-24 A quality cut with oxyacetylene cutting equipment.

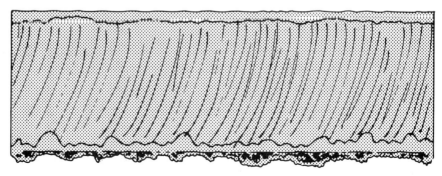

13-25 Cutting speed was too fast.

"Too slow" cut

In all probability, your first cut through steel with oxyacetylene will have a buildup of slag on top, on the bottom, and on the face of the cut. You might find that some of the cutlines are straight, while in other areas no cutlines are visible. All this is largely due to heating the metal too much. In other words, you could have moved the cutting torch quicker along the cutline.

Another characteristic of moving the cutting torch too slowly might be burn holes in the top of the cut. If slow movement is combined with an irregular torch movement, you might end up with quite an ugly edge to the cut metal. This will appear as bubblelike mounds and very uneven cutlines on the face of the metal (FIG. 13-26).

"Too much preheat" cut

If you apply more heat than is required to the cutline before depressing the oxygen cutting lever, the top of the metal will indent. This happens because more the metal along the cutline is ready for the chemical reaction of the oxygen. In extreme cases, the top edge of the cut will appear to flow down the face of the cut. In effect, this is exactly what has happened (FIG. 13-27).

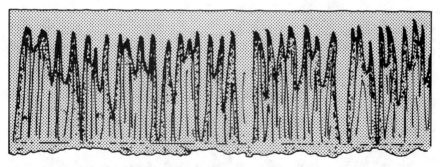

I3-26   Cutting speed was too slow.

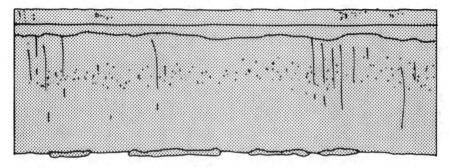

I3-27   Preheat was too hot.

PROFESSIONAL CUTTING TIPS

You should have much control as possible over height and intensity of the cutting-torch preheat flames. The following sections provide a few professional tips to help you make clean cuts.

Observe the metal

One of the most important things that can be learned by beginning welders (in this case beginning oxyacetylene cutters) is to keep a very close eye on the metal. As metal heats up to about the right temperature, you should be ready to depress the oxygen cutting lever and introduce excess oxygen to the cut area. The novice is really at a disadvantage here even if he or she has a keen eye. Until you have watched metal heat up with an oxyacetylene flame a few times, you will not really know what to look for. Be aware that metal rises in temperature quite rapidly and changes from gray to red quickly and then passes through several shades of red until it is practically white in color. If heat is constantly applied, the metal will turn white and become truly molten, just as when you were puddling metal during the welding practice exercises.

When you have allowed the metal to melt without first adding pure oxygen, it will not cut well. When the oxygen is introduced at this point, the metal will deteriorate very rapidly. The time to depress the oxygen-cutting lever is just before the white-hot stage (FIG. 13-28); only then can you expect to make clean cuts.

Adding oxygen

The manner in which you depress the oxygen-cutting lever will have a very real effect on hot the metal cuts, and even on how the metal will look after being cut. It is a common fault of beginners to add too much oxygen when cutting. This is understandable but not acceptable.

The proper way to add oxygen to cherry-red metal is to press very slowly on the oxygen-cutting lever until the metal is flowing away at an even stream. This is actually a very soft, subtle touch that can only be developed over time. In time, however, you will learn that only a bit of oxygen is needed, and then you will develop the soft touch necessary to make a good clean cut in any thickness of metal.

Torch movement

Keeping the cutting torch moving along a straight line, while keeping the preheat flames at the same height above the metal is another requirement for a good metal cut. As I mentioned earlier, you should use your left hand (if you are right-handed) as a steadying point for the torch while controlling the oxygen cutting lever with your right hand and thumb. This is an awkward position when you have to move the torch to cut along a straight line. There will be a natural tendency to put a curve in the cut. One way to overcome this is to move your left forearm and right hand along the

13-28 You must depress the oxygen cutting lever just before the metal reaches the white-hot stage.

line. The task will be easier to accomplish if you also move your whole body at the same time. A good worktable can help you to move in front of the work while cutting with oxyacetylene (FIG. 13-29).

One way to eliminate need for body movement to make a straight cut is to attach a temporary guide for the cutting torch. You can then simply rest the barrel of the cutting torch on the guide. Move the torch along the cutline, both at the right height and along a straight path.

A guide for any oxyacetylene cutting torch can be fashioned quite simply from pieces of scrap steel, such as a piece of angle iron (FIG. 13-30). The only requirements of the metal used for the guide are that it be straight, at least as long as the width of the cut being made, and—probably most important—the right height for the torch being used.

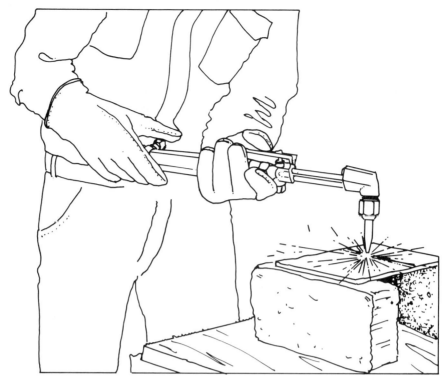

13-29 Lock your elbows against your sides and move your whole body for a straight cut.

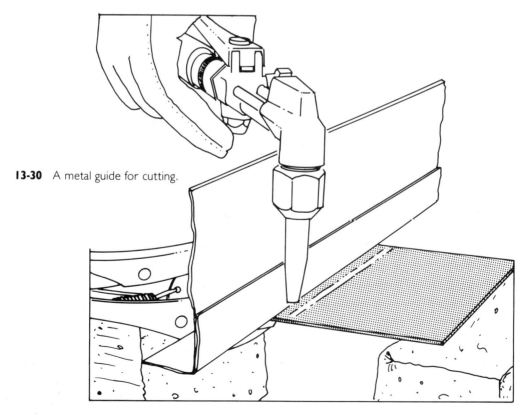

13-30 A metal guide for cutting.

Preheat flame height

As you are probably aware, the preheat flames should be about 1/8 to 1/4 inch above the surface of the cutline. If the white cones are too high above the work, the metal will not heat properly or will take too long. On the other hand, if the heat cones are too close to the work, you will heat up the surface of the metal too much, possibly clogging the holes on the face of the cutting tip.

Setting up a guide

Obviously, some experimentation is called for on your part when setting up a guide for your cutting torch. Once you have found an angle iron that is a suitable height for the torch and tip you are using, you should be able to obtain good results.

The guide, if it is angle iron, can be quickly fastened in place parallel to the cutline with clamps or Vise-Grips (FIG. 13-31). When positioning the guide, do not place it too close to the area to be cut or the clamp might be damaged or fused to the piece. On the other hand, do not place the clamp too far away from the cutline or it will be ineffectual. Most experts agree that a distance of 3/4 inch is about the best distance for standard cutting with oxyacetylene.

A guide enables you to make straight cuts on metal. You must still keep a close eye on the metal so you will know when to depress the cutting lever, however. You will also have to proceed along the cutline at a speed that is adequate for the thickness of the metal being cut.

You should practice cutting straight lines in 1/4-inch-thick steel plate until you can make perfect cuts every time. In the beginning, use some type of guide for the torch. This will enable you to develop a technique

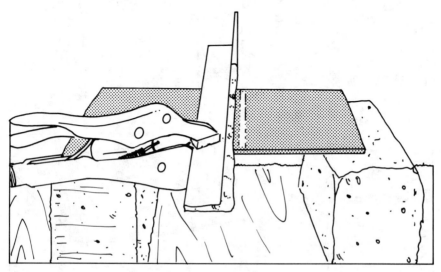

13-31   The cutting guide is simply clamped along the cutline with a pair of Vise-Grips.

with a little more ease. After you have mastered this, try cutting without the aid of a guide for the torch. In time, you should become proficient at cutting freehand.

CUTTING THIN METAL

Thin metal can also be cut with an oxyacetylene cutting torch. Thin metal is any type less than 1/4 inch thick—it is most commonly referred to as sheet metal. For cutting sheet metal, a different approach is needed as there is always a tendency to burn holes in the metal. Two professional pointers will be of use to you when cutting thin metal.

Hold the torch at an angle to the cutline on the surface. A 45-degree angle, or less, will enable you to make the cut without leaving large holes in the cutline (FIG. 13-32). Another point to keep in mind is to work quickly. The more time you spend heating the thin metal, the more chances you have of burning holes rather than cutting a straight line.

When cutting sheet metal, you should use the smallest possible cutting tip. Some experts can make cuts in thin metal with a large welding tip and a slightly oxidizing flame (FIG. 13-33). In time, you may want to experiment with cutting with a welding tip also. For now, however, use a small cutting tip.

BEVEL CUTTING

As you might recall, steel that is thicker than 1/4 inch must have beveled edges or it will be difficult to weld with the oxyacetylene process. Chapter 10 explains grinding the edges of metal to develop a bevel. You might have already tried this and discovered that it takes quite a bit of grinding to bevel the edges of steel plate. Another way to bevel edges of plate steel is, of course, to cut a bevel with a cutting torch (FIG. 13-34). While this type of cutting is not difficult, it does require a slightly different approach from that used for straight cutting.

The major difference between straight cutting of steel plate and bevel cutting is the angle at which the cutting torch is held. To cut a bevel, the

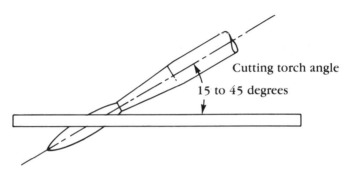

13-32 Cutting thin sheet metal with an oxyacetylene cutting torch. The torch angle is slight to moderate.

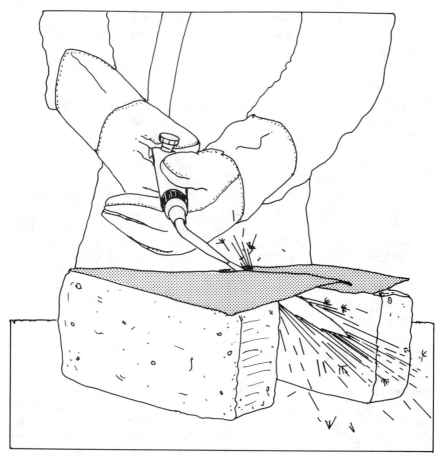

13-33 Light sheet steel can be cut with a welding torch. A very slight angle is used for the torch.

torch tip must be held at a 45-degree angle to the cutline rather than perpendicular (FIG. 13-35). In most cases, some type of guide for the cutting torch will enable you to cut both at a constant bevel and along a straight line.

Another helpful tip for cutting a bevel on steel plate is to use a slightly oxidizing flame. To refresh your memory, first develop a neutral flame for the preheat holes and then add a bit more oxygen to the flames.

It is important, when making a bevel cut, to keep the torch moving at a consistent speed. An unsteady cutting speed will often result in an irregular cut and possibly a stop in the cutting action.

CUTTING THICK STEEL

Cutting thick steel can also be done with an oxyacetylene cutting torch. Generally, the thicker the metal, the longer it will take to make a cut. The

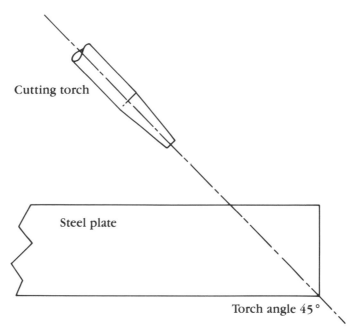

Cutting torch

Steel plate

Torch angle 45°

13-34 Cutting a bevel edge on steel plate with an oxyacetylene cutting torch.

13-35 The cutting torch is held at a 45-degree angle to the cutline to make a bevel cut.

cutting operation is basically the same as those previously discussed, with a few modifications. One of the first things to keep in mind is that the cutting torch tip must be held closer to the work surface of the metal. In most cases, this means 1/6 to 1/8 inch above the cutline. It is also important that the torch tip be held at a right angle to the cutline at all times. This will ensure that the heat is being applied evenly as the cut progresses (FIG. 13-36).

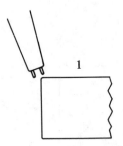

1. Start to preheat; point tip at angle on edge of plate

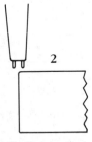

2. Rotate tip to upright position

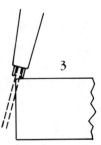

3. Press oxy valve slowly as cut starts, rotate tip backward slightly

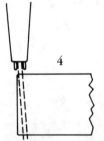

4. Now rotate to upright position without moving tip forward

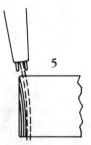

5. Rotate tip more to point slightly in direction of cut

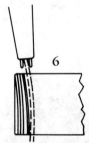

6. Advance as fast as good cutting action will permit

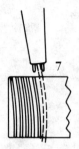

7. Do not jerk; maintain slightly leading angle toward direction

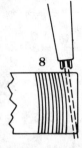

8. Slow down; let cutting stream sever corner edge at bottom

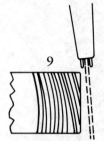

9. Continue steady forward motion until tip has cleared end

13-36   Recommended procedure for efficient flame cutting of steel plate.

Probably the most difficult part about cutting thick metal with oxy-acetylene is starting the cut. There are several different ways of starting a cut that are used by professional welders. One method, assuming you are right-handed, is to begin at the edge of the metal and move from left to right. This provides a clear view of the cut and allows you to look into it to make sure the slag is being blown out by the oxygen. When moving from left to right, you might have some difficulty in following the cutline. You can usually overcome this by using some type of guide for the cutting tip.

Another method used to start a cut in thick metal is to begin at the corner of the metal by slanting the torch in a direction opposite of the direction of travel. As the corner heats up to cherry-red and is cut, move the torch to a vertical position until the steel has been cut.. Then proceed along the cutline.

Some professional welders like to chip the edge of the cutline with a cold chisel and hammer before starting the cutting operation (FIG. 13-37). The sharp edges of the chip will preheat and oxidize more rapidly.

One last method of starting a cut involves using an iron-filler rod. The rod is placed along the cutline before starting (FIG. 13-38). Then, as the preheat flames are applied, the filler rod will reach the cherry-red stage very quickly. As soon as this happens, the oxygen cutting lever is depressed, causing the filler rod to oxidize. A chain reaction then takes place, and the parent metal begins to oxidize at about the same time.

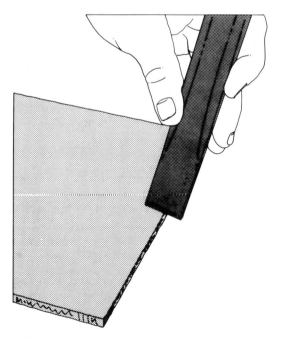

I3-37 Some professionals like to chip the edge of thick steel plate before starting the cut.

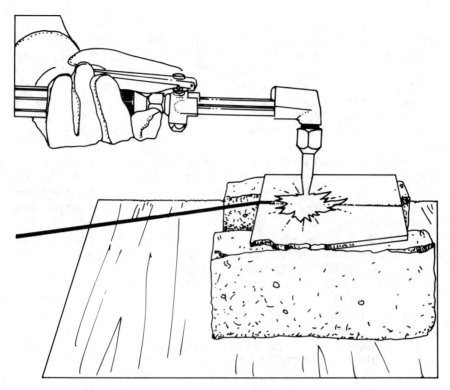

13-38 A welding rod is used here as a cutting guide and aid.

Whichever means you use for starting a cut, and you should try them all, it is important to move the cutting torch at a steady rate once you begin the cut. After the edge of the metal has been cut, keep the oxygen cutting lever constantly open by pressing it lightly. As the metal heats up along the cutline, more cutting oxygen is added. At the same time, move the torch slowly and steadily along the cutline.

PIPE CUTTING

One last use for the oxyacetylene cutting torch is cutting pipe. After all, how can you possibly expect to go to Alaska to work on the pipeline as a welder if you have no idea of how pipe is cut. Think of all the salmon fishing you would miss!

Pipe marking

Before metal pipe can be cut, it must be marked with a soapstone pencil and a special marking guide. Any welding supply house will sell special pipe-marking paper, but you can usually find something around the shop that will do the job just as well, and it won't cost you anything. What you need is a strip of stiff paper, such as matboard, about 1 to 2 feet long and about 6 inches wide (FIG. 13-39). This paper is then wrapped around the

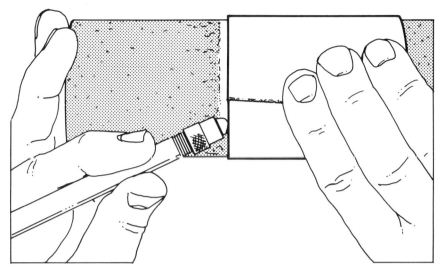

13-39 A piece of thick paper or cardboard, with a straight edge, can be used as a marking guide for pipe.

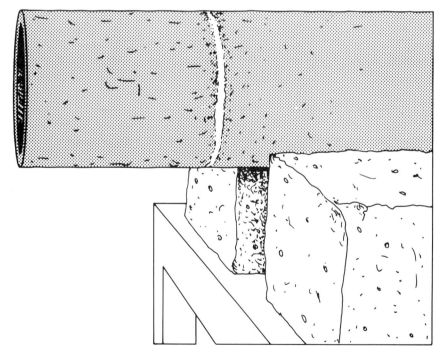

13-40 Steady a length of pipe between two bricks before cutting.

pipe until it overlaps itself. Make sure the edges overlap evenly, and then mark the pipe for cutting.

After the pipe has been clearly marked for cutting, place it on your worktable. Prevent the pipe from rolling off the table by placing it between firebricks (FIG. 13-40). The area to cut should extend off the worktable.

Torch angle

The size of the pipe will determine the angle at which you hold the cutting torch. For pipe that is less than 4 inches in diameter, you should hold the torch so that the preheat flames are almost tangent to the circumference of the pipe (FIG. 13-41). For pipe that is thicker than 4 inches, you can

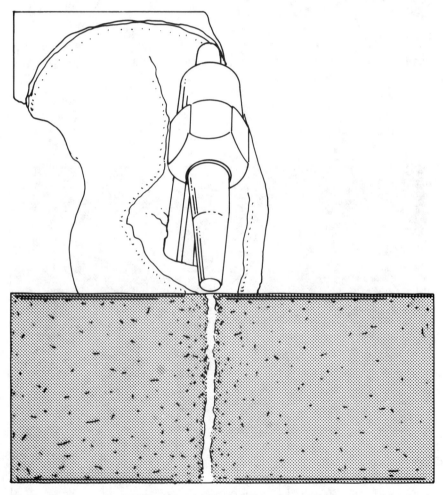

13-41 Cutting small pipe less than 4 inches in diameter requires only a slight torch angle.

usually hold the torch perpendicular to the pipe without danger of burning through the other side of the pipe (FIG. 13-42).

The thickness of the walls of the pipe, rather than the diameter of the pipe, will determine the size cutting tip to use in the torch. Consult the guidebook that came with your torch for the proper tip-size for the pipe you are cutting.

Two-person operation

Begin cutting anywhere along the cutline on the pipe. As the metal reaches the cherry-red stage, press the oxygen cutting lever slowly until you get a good cutting action going. As the actual cutting begins, you must either move the torch along the cutline or have someone else turn the pipe for you. In most instances, cutting pipe is easier if it is a two-per-

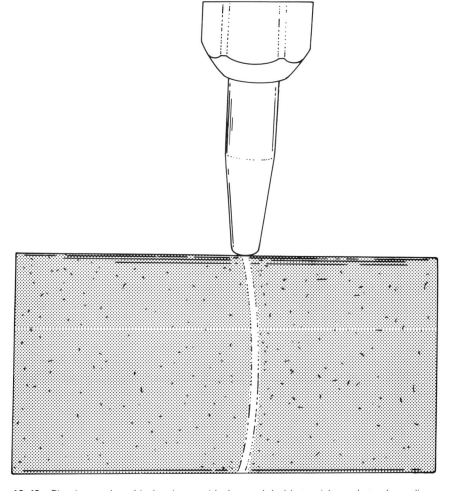

13-42 Pipe larger than 4 inches is cut with the torch held at a right angle to the cutline.

son operation. One person does the actual cutting and the other rotates the pipe, as required. Needless to say, both welder and turner should wear suitable protective clothing and goggles or masks.

Beveling pipe edges

It is also possible to bevel the edges when cutting pipe. This is a bit easier to accomplish if you point the torch towards the short end of the pipe as you proceed with the cutting (FIG. 13-43).

If you find that you must work alone when cutting pipe and the pipe is manageable by one person, you can do the cutting on the ground and use your foot as a means of holding the pipe (FIG. 13-44). Needless to say, you should exercise extreme caution when doing this type of cutting. It is almost as much as balancing act as a cutting operation. Heavy work boots are required.

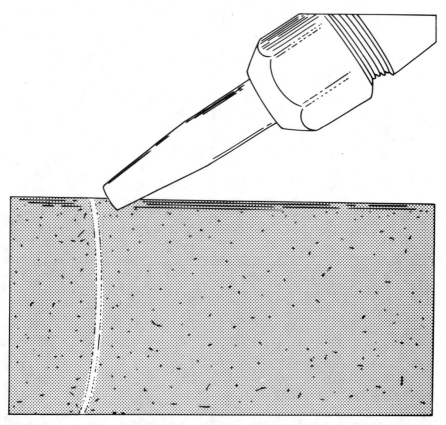

13-43   Torch position for making a bevel cut on pipe. Point the torch towards the short end of the pipe.

13-44   If you are very careful, you can cut pipe in this fashion.

PIERCING A HOLE THROUGH STEEL PLATE

The first exercise in this chapter involved burning a hole in the center of a piece of steel plate. At some time, you might find that you require a hole in a piece of metal. It will be to your distinct advantage to know how to approach the task. The process is really quite simple and can be mastered in a short period of time.

Begin by choosing a cutting tip that is one size larger than that which is recommended for the standard cutting on the particular thickness of metal you are working on. Next, hold the lighted tip of the cutting torch $1/16$ to $1/8$ inch above the surface of the metal. It is important that the tip be held at a perpendicular angle.

Heat the spot until it is cherry-red, and then very slowly depress the oxygen cutting lever. At the same time, the nozzle of the cutting torch should be raised slightly off the surface, say $1/2$ inch. This reduces the chance that slag will be blown up into the nozzle holes and clog them. As the cutting takes place, you might find it necessary to lower the tip slightly and then raise it once again. In a matter of moments, a hole should be blown through the plate. When this happens, lift the torch up and away from the work.

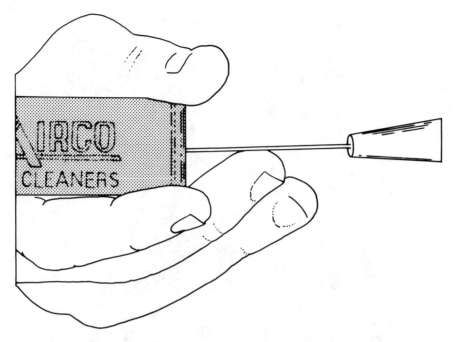

13-45   A clean cutting tip works best.

Piercing a hole in steel plate always runs the risk of clogging the holes in the cutting tip. You should therefore keep a close eye on the flames. If they appear erratic, chances are very good that bits of molten metal or slag are building up on and around both the preheat and oxygen cutting holes. If this happens, you should stop work and clean the tip with a special tip-cleaning tool (FIG. 13-45).

Oxyacetylene cutting is a dramatic experience because of the volume of sparks produced and the speed at which metal can be severed. It will take several hours of practice before you can cut metal well, but it is a welding skill worth knowing. There is something almost magical about working with molten metal, even though there are inherent dangers.

In the next chapter I'll discuss how to accomplish many welding and cutting tasks safely. It will be to your distinct advantage to make safe welding and cutting the only way you know how to work.

Chapter **14**

Safety

From the moment we enter into this world until the last beat of our heart, there will always be some type of danger. Depending on how and where we live our lives, the way we get from one place to another, and how we accomplish various tasks, we always run the risk of injury or worse. That is the natural scheme of things.

The home welder, in addition to all the normal risks to health, lives just a bit closer to danger simply because of the nature of welding equipment. Any welding or cutting operation should never be considered completely safe because to do so would make working with extreme heat second nature, which it surely is not. If you forget for a moment that you are working with a flame that can be as hot as 6,000 degrees Fahrenheit, or that you are around pressurized containers that contain explosive gases, you are flirting with mishap or disaster to yourself and others. While it is impossible to eliminate all the dangers of working with a torch and hot metal, it is possible to reduce the inherent risks by developing safe working habits and religiously following accepted and recommended procedures for handling the equipment.

The purpose of this chapter, then, is to make you aware of the dangers of working with welding equipment, specifically the materials and tools for oxyacetylene welding. With this information in the back of your mind, you should be able to perform all welding tasks with the distinct advantage of knowing how things in general can be done safely. By following guidelines, you will be able to enjoy metal joining and protect yourself from harm.

CLOTHING

Because you cannot control the direction that sparks fly when you are welding or cutting, you must protect your skin from possible damage.

The best way to do this is to wear specially made clothing that will not support combustion. This type of clothing is available but it is quite expensive. Even so, if you plan to do a good deal of welding, it might be to your advantage to invest in fireproof clothing. This is often a requirement for industrial welding (FIG. 14-1).

14-1 Since a welder is always close to the work, it is important that protective clothing be worn.

Cowhide garments

As you might have guessed, there are alternatives for the home welder in the clothing department. Probably the best choices are garments made of thick cowhide (FIG. 14-2). A jacket with long sleeves and a high collar can effectively protect the top half of your body from burns. Leather pants or chaps are a good investment if you plan to do a lot of welding. Although these garments amount to a sizable investment—about $100 to $150 per suit—they are a very good way of protecting your body from the hazards of flying sparks and hot slag.

Cotton and wool apparel

Cotton and wool clothing are undoubtedly the most popular types of clothing for both the casual and professional welder (FIG. 14-3). The ever trusty Levi trousers are heavy enough to shed sparks and will not burst into flames if subjected to an open flame. They will smolder, however, but this does not seem to be a problem because an area can be quickly extinguished with a brush of a gloved hand or a few drops of water.

It is important that all clothing worn during welding be free of any trace of oil or grease. These materials burn violently in the presence of pure oxygen.

Heavy, long-sleeved shirts made of wool are popular with welders. Wool is nonflammable, lightweight, and has a relatively long life. Both cotton and wool clothing can be washed frequently without appreciable wear and tear. Regular and periodic washing of your welding clothes is necessary for safety. Though the material itself might not burn, some forms of dirt or grime will and should be removed as often as necessary.

Flameproof clothing

You can flameproof your welding clothing quite simply by soaking them in a special fire-retardant solution. First, mix up a batch of 12 ounces sodium stannate to 1 gallon of warm water. Second, submerge the clothing in this mixture and let it soak for about 15 minutes. Next, wring out the material and dip it into a solution of 4 ounces of ammonium sulfate to 1 gallon of water. After about 15 minutes, remove the garment from the solution, wring it thoroughly, and let it dry. Clothing treated in this manner will be flameproof and quite safe to wear when welding or cutting.

Flameproof clothing requires special care. Wear the clothing only for welding. Hang clothing in an area where it will remain clean. When clothing becomes dirty, do not wash it; instead, have the clothing dry-cleaned. When waterproof clothing is treated in this manner, it should last for about six cleanings.

Whenever you are wearing clothing that is not specifically designed and made for welding, you should know that cuffs on pant legs and pockets on both pants and shirt are potentially dangerous. These parts of clothing have a tendency to catch and hold hot sparks. Roll down cuffs on pants and remove pockets on your welding shirt or, at the very least, button shirt pockets closed.

14-2 Thick cowhide welding jacket and gloves.

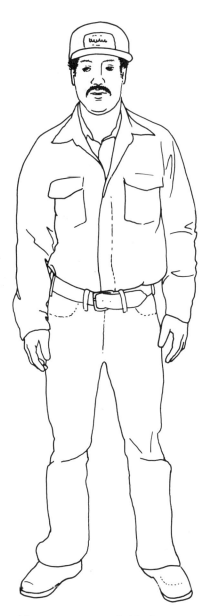

14-3 A hat, wool shirt, and heavy pants are suitable for most types of oxyacetylene welding.

Head covering

Some type of head covering should be worn when welding. A baseball cap or similar hat will not only keep your hair clean, but protect against hot sparks. Welding and cutting are done at less than an arm's length away and there is always the possibility of being struck by sparks. Even though there is little you can do to prevent this, you can and should take precautions to protect ourself from the hazards of red-hot sparks.

A baseball cap is most effective if worn with the bill on the back of the neck rather than in the conventional manner. This offers added protection against a hot spark that could go down your neck. Moreover, with the bill to the rear, it is easier to manipulate welding goggles or a face shield (FIG. 14-4).

14-4 Tinted goggles and a baseball cap are a good combination for head protection during welding.

Boots and shoes

It is important to consider what is worn on the feet. Heavy-duty leather boots are the best choice; those without laces are also at the top of the list. There are any number of boots and shoes designed for welders. Some styles have the added feature of metal toe inserts, which prevent damage to the foot from falling heavy objects.

Lightweight shoes or, heaven forbid, sneakers or tennis shoes should never be worn when welding. Although boots are the best protection, heavy work shoes can also be worn, provided the pant legs cover the tops of the shoes. Because you will be concentrating your energy on the area being welded, you might not notice a hot spark burning a hole in the top of your shoe until the damage is done and your foot gets burned.

EYE PROTECTION

Surely one of the cardinal rules of welding is: Never look at a welding flame or arc unless you have suitable protection covering your eyes. Eye protection is necessary not only to prevent damage to the eye from the injurious rays created during welding or cutting, but also to protect the eyes from flying sparks and hot slag.

Goggles

There is a wide selection of good eye protection available to anyone planning to weld. In fact, most welding outfits will come with a pair of special goggles in the set. Goggles are available in both high-impact plastic and safety glass. Given the choice, most professionals will choose the safety glass type. These goggles do not scratch, and tend to offer an undistorted view of the work. Any quality pair of safety goggles has lenses that can be removed and replaced if the need arises (FIGS. 14-5 and 14-6).

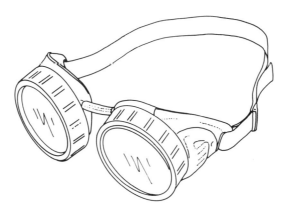

14-5 Light-duty welding goggles.

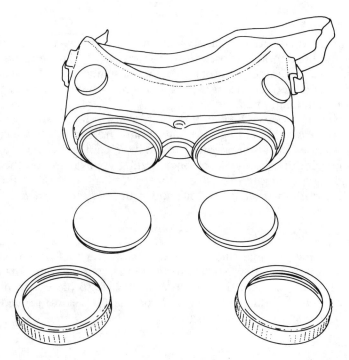

14-6 Quality welding goggles will have extra tinted glass lenses to replace the original lenses.

Face shields

Also available are face shields in both clear and tinted styles. Face shields have a few advantages over goggles. The field of view is greater when the welder is wearing a face shield, enabling the welder to move about the work area more naturally. When wearing goggles, however, the welder might feel quite comfortable while working on a project in a fixed location, but find moving around the workshop difficult. Movement with welding goggles gives the distinct feeling of tunnel vision.

A face shield offers more protection than a pair of goggles simply because it covers the entire face and part of the head (FIG. 14-7). Every face shield I have ever seen could be easily lifted up and away from the face, offering the chance for quick, normal vision if the need should arise. A pair of welding goggles, on the other hand, might not be quite so easy to remove quickly, especially when a hat or helmet is worn along with them.

One of the disadvantages of face shields is that they are all made from a plastic material, very much the same as a face shield for a motorcycle helmet. Plastic can scratch easily, and the welder's field of vision might become distorted somewhat. The only practical solution is to replace the lens with a new one.

14-7   A tinted face shield designed for oxyacetylene welding and cutting.

Filtering lens

It is a fallacy to think that any pair of tinted glasses or goggles can be used for all types of welding or cutting. A special lens is needed to filter out harmful rays from welding. This filter must be of a type that is suitable for the particular welding or cutting task at hand. A filter's job is to absorb ultraviolet and infrared rays as well as most of the visible light rays. It is important, then, that you use only goggles or other eye protection that has been specifically designed and approved for welding or cutting.

If you wear corrective lenses, speak to your optometrist about buying a special pair of welding goggles tailored to your particular vision requirements. Also available are tinted welding goggles that fit over conventional eyeglasses.

Clear glass

In addition to a pair of tinted welding goggles, the home welder should have some type of clear eye protection. In most cases, a clear face shield will be more than adequate. To prevent eye damage, use clear eye protection gear when grinding metal (FIG. 14-8). Never rely solely on the shields that are common on most quality bench grinders as they will do little, if anything, to protect your eyes from flying sparks and metal particles.

It is a sound principle to keep your clear eye protection gear right next to your bench grinder or other abrasive machinery. Then get into the habit of putting on your eye protective gear before turning on the machine. This small act will go a long way in protecting your eyes from damage.

RESPIRATORS

Occasionally a welder might come in contact with fumes (both nontoxic and toxic) as well as normal air that is heavy in particulate matter. Probably such contact occurs most often when the welder is grinding metal on a bench grinder, wire brushing metal, or working with a disc sander.

14-8 Wear a clear face shield when working on a bench grinder.

While in most cases this "heavy air" will do little harm to the welder, some measures should be taken to protect the respiratory system.

A number of different types of respirators are available. The range goes from a bandana secured over the nose and mouth to self-contained breathing masks that would be quite effective during a poisonous gas attack. A little common sense should come into play here. For example, if you are grinding metal that has a lot of surface rust and you see that you will be working on the piece for a long period of time—more than a few minutes—you should wear an adequate covering over your mouth and nose.

One of the more popular means of protecting the respiratory system is an inexpensive cloth mask (FIG. 14-9). You can pick one of these masks up at most welding supply houses. They are also commonly sold in paint stores and lumberyards. Since their cost is minimal at the present time— about three for one dollar—you should pick up a supply and keep them handy around your welding and workshop. Get into the habit of using a respirator whenever you must work in an area that has a heavy concentration of particulate matter in the atmosphere. This will reduce the chances of serious respiratory damage as a result of breathing this type of air.

14-9 Wearing a respirator is a good precaution when sanding or wire-brushing rusted metal.

GLOVES

Another cardinal rule of welding is: never pick up a welding blowpipe or cutting torch without a pair of heavy-duty gloves on your hands. By following this simple rule, you will dramatically reduce the chances of burning your hands.

Since at least one of your hands will always be within 1 foot away from a flame with a temperature of up to 6,000 degrees Fahrenheit, you should always protect your hands from sparks and reflected heat. The best way to do this is to wear a heavy-duty glove on each hand (FIG. 14-10).

There are easily hundreds of different types and styles of gloves designed for welding. Add to this amount a vast selection of general-purpose heavy-duty work gloves, and you will have little problem in finding a pair of gloves that will effectively protect you while you are welding or cutting. It is important to keep in mind, however, that not just any pair of gloves is suitable for working around hot metal.

The gloves you wear for welding can be a bit lighter than those you wear for cutting. But, in all cases, the gloves should be heavy and practical. There is a cutoff point. A pair of gloves that are more than adequate for working with a torch might also be too heavy for effectively controlling the flow of oxygen or acetylene. In other words, the gloves might not allow you to turn the control knobs on the blowpipe handle. Consider this carefully when shopping for a good pair of gloves. Purchase a pair that will offer adequate protection from the heat and will also allow you to adjust equipment controls properly with your fingers.

14-10 A welder's hands are always close to the heat, so a good pair of gloves is very important.

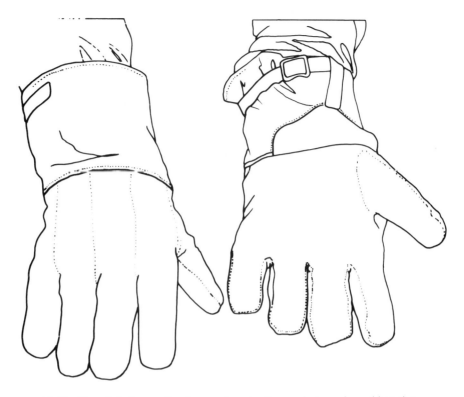

14-11 Gauntlet gloves offer the most protection against sparks and hot slag.

All experts agree that a pair of welding gloves should be flameproof and of the gauntlet type (FIG. 14-11). You will have no problem finding this longer type of gloves at any welding supply house. At the present time, welding gloves are made from heavy-duty leather, special fire-retardant materials, or a combination of both.

Personally, I have found that a pair of welding gloves are easier to work in if they are made from leather. Leather, because it is a natural material, tends to facilitate hand and finger movement a bit better than man-made fibers. Many styles and thicknesses of welding gloves are made from leather. Generally speaking, leather welding gloves do not insulate as well as fire-resistant gloves. Therefore, the latter might be a better choice if you work for a long period of time around molten metal. Keep in mind, however, that certain fireproof gloves tend to be stiffer than leather gloves. As a result, there will be a loss of precise finger movement. As I mentioned earlier, there always seems to be some type of trade-off between gloves that will totally protect your hands and gloves that facilitate finger movement. In the end, the choice must be yours. You will probably end up having at least one pair of leather and one pair made of some fire-retardant fabric.

APRONS

One last piece of clothing that is worth mentioning is an apron for welding. A quality apron, designed specifically for use during welding, offers one very good alternative to traditional heavy-duty clothing. Aprons are currently available in many different styles, materials, and price ranges. To be effective, the apron should protect as much of the welder's front as possible while enabling the welder to have free movement of arms and body.

Leather and fire-retardant, or heat-resistant, fabrics are probably the two most common materials used to make welding aprons. Both materials are quite effective at reflecting heat and defecting hot sparks while someone is welding and/or cutting. Look at the aprons offered at your local welding supply house. You quite possibly will find one suitable for your welding protection needs. Obviously, all the requirements mentioned earlier for the care and use of clothing apply to aprons designed for welding (FIG. 14-12).

FIRES AND FIRE EQUIPMENT

Because the possibility of a fire always exists for the welder, he should have some means of putting a fire out if one should start. Before you can put out a fire, you must know a few things about fires in general. The old military saying, "Know your enemy," is one of the most effective ways of dealing with fires. As a way of making you more aware of fires in general, we need to discuss the basic components of fires.

Fuel, oxygen, and heat

Before a fire can take place, three things are needed: fuel, oxygen, and heat. If any of these three components is removed, a fire cannot continue to burn. Whenever a fire breaks out, you must make a quick decision

14-12 A bib apron provides adequate protection of clothing.

which of the three parts of that fire can be eliminated most easily in order to eliminate the fire. The three most obvious means of doing this are removing the fuel (which can often be nearly impossible), cooling the material that is burning, or nearly removing the oxygen by smothering.

All fires have two things in common, heat and oxygen, and these can be an aid to you in fighting any fire. The third component of a fire, namely the fuel, can vary from one fire to another. It is because of the possible variation of fuel in any given fire that all fires are grouped into four possible types: Classes A, B, C, or D. You should know about each class of fire in order to determine the best way to extinguish any particular fire.

Class A fires

Class A fires are those in which wood, paper, clothing, and similar materials are burning. Generally speaking, water is the best and most effective way to fight a Class A fire. Water will penetrate into the burning material and cool it. Water will also, at least in part, seal the material from the oxygen in the atmosphere, although to a much lesser extent than other fire-fighting materials.

Class B fires

Class B fires consist of flammable liquids such as oil, gasoline, paint, and similar materials. As a rule, once the flame is extinguished in a Class B fire, the fire will go out. You must, in effect, remove the oxygen or separate the burning material from oxygen to extinguish a Class B fire. Only the vapor on top of the liquid surface is burning and not the material itself in a Class B fire. For example, diesel fuel burns on the surface as the vapors evaporate into the surrounding atmosphere. The fuel itself does not contain oxygen. Separate these fumes from the oxygen in the atmosphere, and the fire will go out.

The most effective means of fighting a Class B fire is, of course, to smother the fire, thereby separating the vapor, which is burning, from the fuel or burning material, which is not burning. Foam, powder, sand, and other types of nonflammable materials are most effective because of the heavy smothering action. Once the burning material is separated from the oxygen in the atmosphere, the fire will go out almost immediately.

In no case should water ever be used to extinguish a Class B fire. Most flammable liquids will float on top of water and continue to burn. Water will not, therefore, extinguish a Class B fire. In fact, when water is thrown on a Class B fire, it might actually float the flames and fuel over a large area, which could dramatically spread the fire and prevent putting it out.

Class C fires

Class C fires are electrical in nature. This class of fires includes electrical wiring, motors, switches, machinery, and appliances. A smothering action is the best means of fighting a Class C fire after the power has been turned off. For our purposes, that means a nonconductive material, generally a special type of fire extinguisher. Because of the possibility of electrical shock, water should never be used in an attempt to extinguish an electrical fire.

Class D fires

Class D fires are rather rare and almost nonexistent for the home welder. These fires are caused by magnesium shavings. The only way to put out a Class D fire is to use a specially designed fire extinguisher specifically labeled for Class D fires. If the home welder ever plans to work with combustible metals, he should have a fire extinguisher of this type handy.

Any fire will fall into a least one of the above classes of fires: A, B, C, or D. Probably the best way to fight and extinguish a given fire is to use a fire extinguisher that has been designed to put out that specific class of fire.

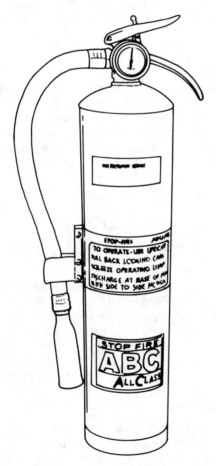

14-13 This fire extinguisher is designed to fight Class A, B, and C fires.

Fire extinguishers

As you might have guessed, there are fire extinguishers for each class of fire. There are also fire extinguishers that are suitable for more than one class of fire. These will be clearly labeled. Probably the best choice for the home welder is a type generally referred to as multipurpose or all-purpose fire extinguishers. These are clearly labeled for "ABC Fires" (FIG. 14-13).

The best types of fire extinguishers will have a pressure gauge at the top of the unit, which will indicate the condition of that particular fire extinguisher. Some are more elaborate than others, but all that is really required is a gauge that will indicate whether or not the unit will work when needed. A gauge like this will give you valuable information at a glance: *recharge* or *operable*, for example (FIG. 14-14).

Even though your fire extinguisher might have a gauge, you should have it checked by a qualified repairman once a year. This will ensure that the unit is, in fact, in good condition and that it will be an effective means of fighting a fire.

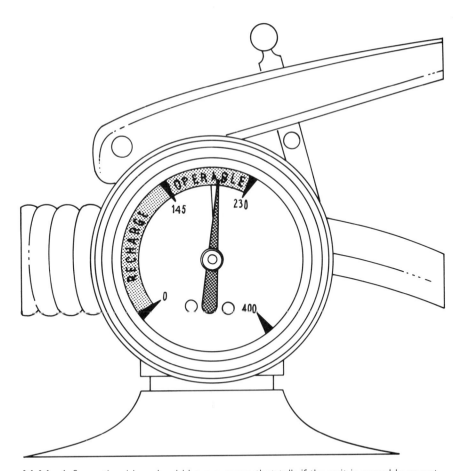

14-14 A fire extinguisher should have a gauge that tells if the unit is operable or not.

If you ever use a fire extinguisher, you should have it refilled, even if you only use it for a few seconds. What would be the point of having a fire extinguisher around your shop if it was only half full or not properly charged? Your property and possibly your life might depend on your ability to extinguish a fire quickly. This is reason enough to make certain your fire extinguisher is in first-class working order at all times.

Sand and water

In addition to a quality and multipurpose fire extinguisher, you should have a bucket of dry sand and a pair of water buckets in your workshop or in the area where you are working. Sand can be used quite often to extinguish a Class B fire because it will quickly smother the flames. For the reasons previously mentioned, a bucket of water should be used only for putting out a Class A fire. Nevertheless, both sand and water can be used to quickly extinguish certain types of fires and offer alternatives to your fire extinguisher (FIG. 14-15).

Obviously, your best defense against fires is to prevent them from happening. That means working only in safe surroundings. Since the next chapter covers many aspects of the home welding shop, you will have to turn there for more information on the best type of surroundings in which to weld or cut.

BURNS

Since we are discussing fires, it will be worthwhile to mention a few things about burns. To say the least, burns resulting from welding or cutting are not a pleasant topic for discussion. Nevertheless, the home welder should know what to do in the event of a serious burn.

Generally speaking, all burns can be grouped into three rather large categories. Since treatment of any one particular burn will depend on the nature of that burn, we will discuss all types of burns. All burns fall into either first-, second-, or third-degree burn categories.

14-15 Sand and water can be used to fight certain types of fires.

First-degree burns

First-degree burns result in red skin and possibly a small amount of pain. Sunburn is usually a first-degree burn. The same is true of red skin due to reflected heat from welding or cutting. Generally speaking, first-degree burns are less serious than they are uncomfortable. The best treatment for first-degree burns is to relieve the slight pain by applying a burn ointment to the area. If the burn covers a large area, see a doctor as soon as possible. Generally, the pain will subside in a few hours. If pain persists, seek medical help.

Second-degree burns

Second-degree burns are a bit more serious and are characterized by blisters. Treatment for second-degree burns should *not* include covering the blisters with a burn ointment. You should prevent the blisters from breaking by covering the area with sterile gauze and holding it in place with a bandage. Second-degree burns are serious, and medical attention should be given by qualified personnel as soon as time permits. The greater the area of the second-degree burn, the more potentially serious the problem.

Third-degree burns

Third-degree burns are the most serious of all burns and are characterized by burned or charred flesh. Because of the nature of welding and cutting equipment, third-degree burns are common. It will therefore be to your advantage to know what to do if this type of burn should happen to you or someone else in the welding shop. Know in advance that third-degree burns that cover more than 50 percent of the body often result in a fatality. Therefore, quick action on your part can make a difference.

There are a few cardinal rules that should be followed when dealing with third-degree burns. They are as follows:

- Do not move the person unless absolutely necessary.
- Do not apply burn creams or salves to cover the burn.
- Do not remove burned clothing. Leave it in place to be removed by qualified medical personnel.
- Treat for shock by covering the victim with a blanket or other suitable material. Lay the person down with feet elevated, if this is practical.
- Seek medical attention as quickly as possible, by either rushing the victim to the hospital or calling in emergency medical assistance.

All burns require attention as they are all potentially serious. It is important to keep in mind that quick action on your part can make a difference, so act accordingly.

HANDLING WELDING EQUIPMENT

The equipment used in welding and cutting is potentially dangerous and should always be treated with respect. Failure to do so might result in mishap. Many points for safe operation can be found throughout this book. The majority of the remainder of this chapter will summarize those practices that should be standard operating procedures for anyone working around or with welding equipment.

Oxygen and acetylene cylinders, as well as other fuel gas containers, have properties that can cause serious accidents, injuries, and even death if proper precautions and safety practices are not followed. It is therefore very important that you become aware of these safety precautions. Make them a part of your welding habits.

Because both oxygen and acetylene are in pressurized containers, they must be handled with respect. It is very important that these cylinders be securely held in an upright position at all times. Probably the best way of doing this is to chain both cylinders to a special hand truck

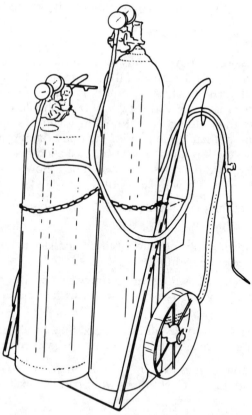

14-16 Oxygen and acetylene cylinders should always be securely fastened so there is no chance of falling. A wheeled handcart is probably the best means of storage for your tanks.

designed to hold oxygen and fuel gas cylinders. These handcarts are available where you purchase your oxygen and acetylene (FIG. 14-16).

If the tanks are to remain in one location, make certain that both are secure by chaining them to an immovable object like an interior wall or column. Never allow either type of cylinder to stand by itself for any period of time. This will be your best insurance against accidental tipping of the cylinder.

Keep all organic materials as well as flammable substances away from both oxygen and acetylene cylinders. Remember that even normal soot and dirt can constitute a combustion hazard around these gases.

Never use either oxygen or acetylene (or other fuel gases) without first attaching an approved regulator to the tank. A full tank of oxygen will have an internal pressure of about 2,200 pounds. Full acetylene cylinders will have an internal pressure of up to 400 pounds. It is therefore extremely important that only a suitable regulator be used to reduce these pressures to recommended working pressures in the lines (FIG. 14-17).

Handling oxygen

In addition to the general rules for safe handling of both oxygen and acetylene, there are other recommendations that apply specifically to oxygen. Never allow oxygen to come in contact with oil, grease, kerosene, paint, tar, coal dust, or dirt. Oxygen supports combustion and can burn violently in the presence of these and other flammable substances. You

14-17 Always use the proper regulator for oxygen and fuel gas.

should never handle oxygen or equipment that might come in contact with oxygen—such as the oxygen regulator—with dirty hands, grease-stained gloves, or other soiled clothing.

Before attaching a regulator to the oxygen cylinder valve, wipe the fitting and threads with a clean, dry cloth (FIG. 14-18). You should also crack the valve on the oxygen cylinder to blow out any dirt or other foreign matter that might have found its way into the valve. When cracking the oxygen cylinder valve, do so quickly. Position your body so that the oxygen in the tank will not spray on you or your clothing. Stand to the side of the tank valve (FIG. 14-19).

Never use oxygen for ventilation or as a substitute for compressed air. You should never blow dust off anything with oxygen. Cleaning your clothes off with a blast of oxygen and then lighting up a smoke is one of the quickest ways to end up in the hospital.

Always refer to oxygen as "oxygen" and never "air." This will avoid the possibility of confusing oxygen used in the welding process with compressed air.

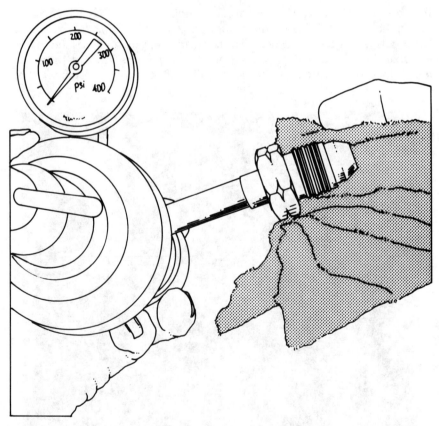

14-18 Always wipe the threads on regulator fittings with a clean dry cloth before attaching to the fuel or oxygen cylinder.

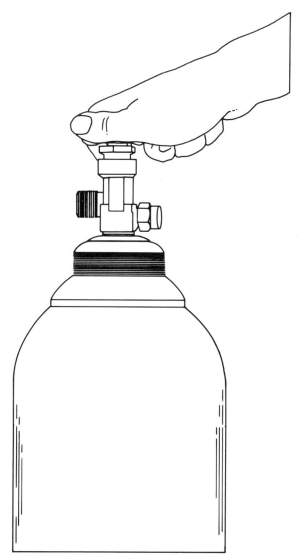

14-19 Stand to the side when you crack an oxygen or acetylene cylinder valve.

Never use oxygen for anything other than its intended purpose. Oxygen that is designed to be used for medical purposes such as therapy for respiratory ailments will have U.S.P. (United States Pharmacopeia) stamped on the cylinder. All other types of oxygen are for industrial uses only and should not be used for oxygen therapy.

When transporting oxygen, do so only with the container in an upright position in an open vehicle. Never carry an oxygen cylinder lying on its side in the closed trunk of an automobile.

Handling acetylene

Acetylene cylinders are equally dangerous. All the guidelines for safe operation and use of oxygen cylinders should be followed for this fuel gas as well. In addition, you might recall that all acetylene cylinders have a safety plug in either the top or bottom of the cylinder. This safety plug is designed to melt at a temperature of about 212 degrees Fahrenheit. It is a safety feature intended to reduce the hazards of an explosion in the event of a fire (FIG. 14-20). Because this safety plug has such a low melting point, you should never work with an open flame around a cylinder of acetylene. For the same reason, you should never pour boiling water on top of an acetylene vale in an effort to thaw out a frozen valve. This can happen when an acetylene cylinder has been stored in an unheated room or outdoors during cold weather. At all times you should protect the fusible safety plug in acetylene cylinders from damage. In addition, always store, transport, and use acetylene in a vertical position.

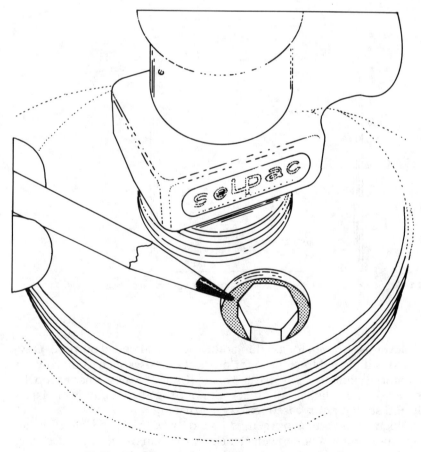

14-20 The safety plug in the top of an acetylene tank.

REVIEW OF EQUIPMENT SETUP

When setting up welding equipment, specifically oxygen and acetylene cylinders, follow those guidelines set down in chapter 1. A quick review of those steps might further drive home the importance of proper handling of all welding equipment.

Oxygen connections

Make certain that the fuel and oxygen cylinders are securely held in one place that will permit them to have a safe, happy life. Crack the valve of the oxygen cylinder before attaching the special oxygen regulator. Wipe the threads of the fittings of both the regulator and cylinder valve before fastening the regulator in place.

Tighten the regulator with a wrench designed for the purpose (FIG. 14-21). Undertightening will result in a leak, and overtightening can cause the brass fitting to become strained and weakened.

Connect one end of the oxygen hose (green in color) to the regulator and the other end to the oxygen fitting on the torch handle. Snug up all connections with a suitable-size wrench.

Before opening the valve on top of the oxygen tank, check to make sure the oxygen control knob on the blowpipe handle is closed. In addition, open the regulator control lever until there is no resistance in turning of the lever. The diaphragm is now open. Open the oxygen control valve on the cylinder slowly (FIG. 14-22). This will ensure that the delicate oxygen regulator does not receive a surge of pressure from the cylinder. Remember that a full tank of oxygen will contain approximately 2,200 pounds per square inch of internal pressure. Once a bit of oxygen is in the regulator and the high-pressure gauge registers the pressure in the tank, open the valve by two turns. At this time, give a quick check to the regulator to make certain that it is not leaking. If you suspect it is, use the soapy water test.

In most cases a leak is a result of not tightening a connection enough. If you discover a leak, first turn off the valve on top of the tank. Then try snugging up the connection with the proper-size wrench. Put pressure into the system and check the suspected area (FIG. 14-23). If the leak persists, there might be internal damage to the regulator or fitting. It should therefore not be used until the problem has been sufficiently corrected.

Acetylene connections

The procedure is essentially the same for hooking up acetylene connections. Keep in mind, however, that all connections for acetylene are of the left-handed type and will be marked with a groove cut around the circumference of the brass fitting (FIG. 14-24).

After the recommended working pressures have been set on the regulator, open the torch handle momentarily to see that acetylene is, in fact, flowing through the line. You should also keep one eye on the regulator dials and make sure that they move slightly, if at all.

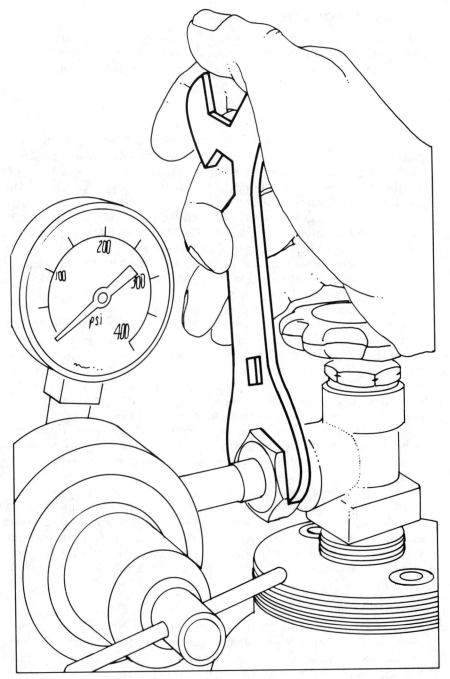

14-21 Tighten the regulator fitting with a suitable-size wrench.

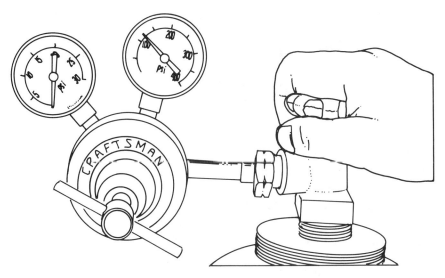

14-22 Always open the cylinder valve slowly so the regulator will not be damaged by a sudden surge of pressure.

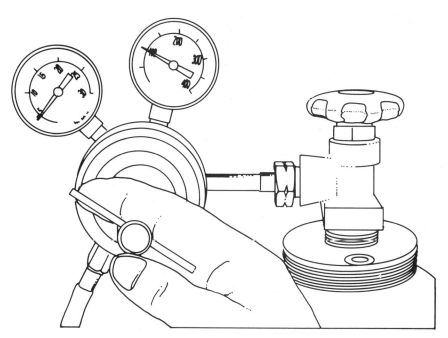

14-23 Set the working pressure by turning the lever until the desired pressure is reached.

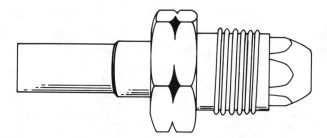

14-24 All fuel gas fittings have left-handed threads and a notch around the fitting nut.

WORKING WITH OXYACETYLENE EQUIPMENT

When working with oxyacetylene equipment, there are a few points to keep in mind. The first is that you should turn off the flow of both oxygen and acetylene at the cylinders if you stop work for more than a few minutes. The second is that if you are shutting down the equipment for more than a half-hour you should, in addition to turning off the flow of oxygen or fuel gas, also bleed the lines. Clear the acetylene line first by opening the control knob on the blowpipe handle. Watch the acetylene regulator when you do this. In a matter of moments both gauges, if so equipped, should register zero. Then turn off the acetylene control knob on the blowpipe handle. After you have cleared the line of acetylene, follow the same procedure for cleaning the oxygen line (FIG. 14-25).

Obviously, when you are shutting down the oxyacetylene equipment, you should make certain that there is no open flame in the area. When working with oxyacetylene equipment, it is important always to have a clear path to the tanks of oxygen and fuel gas. In the event of a mishap, one of the first things you should do is to turn off both oxygen and acetylene at the top of both tanks.

Another good practice when welding is always to make sure that the hoses carrying oxygen and acetylene do not lie under the work. A hot spark or piece of hot slag falling on a hose could easily burn its way through with drastic results.

Store your welding equipment and tanks in a safe place. This equipment, for example, should never be left within easy reach of children.

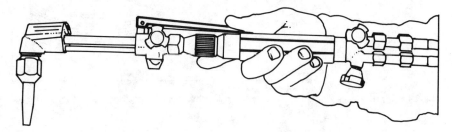

14-25 You can clear the oxygen line by depressing the oxygen cutting lever on the torch. The supply of oxygen must be shut off at the tank first.

Both oxygen and acetylene cylinders come with a special valve protection cover. This cover should be in place over the valve whenever the regulator is not attached. If children have any possibility of being in the area where oxygen or acetlene is stored, the valve protection cap should be made secure and, at the very least, too tight for small hands to loosen (FIG. 14-26).

Hose, torch, cutting attachment, and tips should all be stored in a safe place as well. A cabinet that will keep these parts dust-free is best. It will go a long way to preventing costly repair bills as well as mishaps around the workshop.

WELDING HAZARDS

In summing up this chapter on safety, I think it will be helpful to outline briefly the hazards that are always present in welding. As a welder, you should know these hazards and continually strive to take them into consideration. This will be your best insurance against an unfortunate and possibly tragic experience.

- **Fire**. The possibility of fire always exists. You should therefore always have at least have at least one means of extinguishing a fire if one should break out.

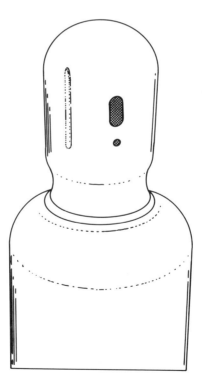

14-26   The safety protection cap should always be on a cylinder of oxygen or acetylene whenever the regulator is not attached.

- **Burns**. Because of the nature of welding equipment, a serious burn can happen at any time. Always wear protective clothing to reduce the chances of burns to yourself or others in the shop.

- **Eye injury**. Always wear tinted eye protection when welding, cutting, or even simply adjusting the flame of your blowpipe. Additionally, you should wear clear eye protection when performing welding-related tasks such as grinding metal surfaces (FIG. 14-27).

- **Explosions**. Always handle, transport, and store oxygen and fuel gas cylinders with the respect due them. Check all connections to make certain that no fuel, gas, or oxygen is leaking into the atmosphere. Exercise care around other containers with flammable materials.

- **Harmful or poisonous gases**. Welding or cutting processes with some types of metals, such as galvanized steel, can produce a harmful gas. Make certain that the welding area is adequately ventilated at all times. Wear a respirator over your mouth and nose whenever grinding or welding potentially dangerous metals.

- **Mishaps**. Unsafe surroundings or unsafe work habits can cause accidents.

- **Failure to exercise caution and common sense**. Think before you act, and you will eliminate 95 percent of the danger associated with welding and cutting operations.

A wise welder knows and practices safe work habits. He is constantly aware of the dangers involved in welding and does everything in his power to make welding and cutting a safe and enjoyable pursuit.

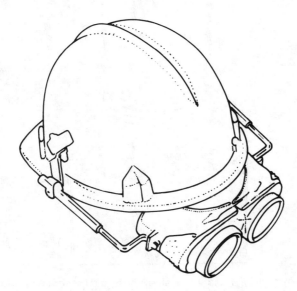

14-27 Always wear eye and head protection when welding or cutting.

Chapter **15**

The home workshop

*I*n addition to oxyacetylene welding equipment—blowpipe, cutting attachments, safety equipment, hoses, tanks of fuel gas and oxygen, and related accessories—the home welder requires a safe space to work. If you think you can take care of all your welding projects in the back of your garage or in your basement, you are flirting with danger. Welding, because of the inherent risks, requires a special area. After all, we are dealing with controlled fire and very hot metal when welding or cutting. If special precautions are not taken, you could have an unfortunate experience at the very least.

OUTDOOR WORKSHOPS

Chances are, you will begin practicing cutting and welding outdoors. This makes sense on a number of levels. Even though there are some drawbacks to working outside, you should find an outdoor workshop more than adequate for your practicing needs.

Ventilation

Probably the greatest advantage of working outdoors is that there is always plenty of ventilation. Even on windless days, smoke and fumes from welding and cutting will quickly drift up and away from the area. As you have perhaps already discovered, the smoke from welding, even though nontoxic, can be upsetting to the system. It is, therefore, a good practice to inhale as little smoke or fumes as possible.

Equipment setup

Setting up equipment, as well as shutting down equipment, can be safely done outdoors as well. For example, when you crack the oxygen or acetylene valves prior to attaching the regulators or close them later, these gases will quickly disperse into the atmosphere and pose little danger. The gases simply drift away. When you are working indoors in a workshop, for example, you must make certain that these gases are exhausted from the area by some artificial means, such as a ventilator fan. Later in the chapter, I discuss the various means of ensuring fresh air, and plenty of it, in an indoor welding workshop.

Minimal cost

Another advantage of an outdoor workshop for welding is that the cost of setting up such a shop will be, in most cases, quite minimal. In addition to the necessary oxyacetylene welding and cutting equipment, you will require a worktable. A worktable is one of the basic components of any welding shop, as well as a real aid to working with oxyacetylene equipment. You will want to build one as soon as you have mastered some of the basic cutting and welding techniques discussed earlier in this book.

BUILDING A WELDING TABLE

To build a simple welding table, you need 20 feet of 2-×-2-inch angle iron, 20 feet of 1-inch-wide steel strap, and 24 8-×-4-×-2-inch firebricks. See TABLE 15-1.

The dimensions for a standard welding table are given in FIG. 15-1. Begin by cutting four pieces of angle iron, each 28 inches long. Then cut the remainder of the angle iron into two, 24-inch lengths and two, 32-inch-long pieces. The angle iron (actually angle steel) comprises both the legs and the frame for the top of this welding table. Cut precisely for best results.

Assembly

Begin assembling the table by laying out the top pieces and welding them together (two 24-inch- and two 32-inch-long pieces). You will be welding a rectangular frame that when finished will measure 24 inches wide by 32

Table 15-1   Materials list for the welding table.

Quantity		Size (inches)
4	28	2″-×-2″ angle iron
2	24	2″-×-2″ angle iron
2	32	2″-×-2″ angle iron
4	24	1″ steel strap
5	32	1″ steel strap
24		8″-×-4″-×-2″ firebricks

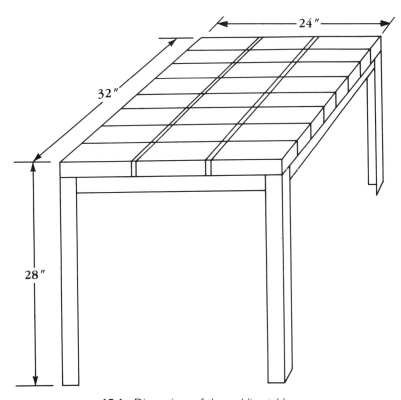

15-1 Dimensions of the welding table.

inches long. Do not overlap the pieces; use the simple butt joint instead. You might find it easier to tack-weld the pieces together before welding. Be sure that the four corners of the table remain square.

After the top frame for the welding table is assembled, attach the 28-inch-long legs. Attach one leg to each corner and tack-weld on the outside of the frame. You now have the basic frame table. All that remains is to weld the steel strap, four 24-inch-long pieces, and five 32-inch-long pieces inside the frame top. Placement is important because the steel strap provides support for the firebricks, which make up the working surface of the table.

It is probably easiest to begin by welding the 24-inch-long steel strap. Space the strap at 8-inch intervals and tack-weld them in place. Next, the 32-inch-long strap is either laid on top of the cross members or woven through them for additional strength. Just the ends of these longer strips are tack-welded until all straps are in place. Then you can go back and tack-weld all joints where straps overlap. Place the longer straps at 4-inch intervals so the edges of the firebrick will lie on them. If you are in doubt as to the placement of the steel strap, simply lay a few firebricks into the frame to give you a clear picture of the exact layout of the straps (FIG. 15-2).

As you can see in FIG. 15-1, the worktable has a finished height of 28 inches. If you are taller or shorter than average, you might want the table

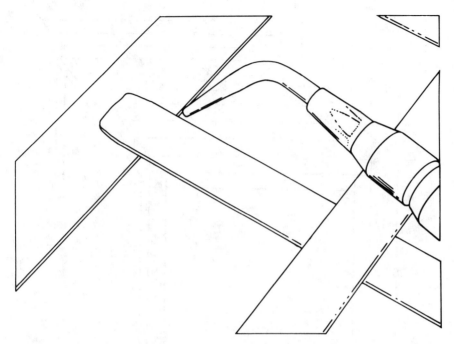

15-2 Tack-weld the steel strap around the inside of the table frame. Proper spacing is
very important.

at a different height. To do this, simply adjust the length of the legs. It
might be of interest to you to know that you will have some extra 2-inch
angle iron if you buy the listed 20 feet. Actually, you will have an excess of
about 16 inches, so you will have little problem increasing the basic
height of this table by about 3 inches with the materials you have on
hand. Twenty-eight inches is, by the way, the standard height for a work
surface, and most people find this height comfortable.

Modifications

You can, of course, modify the basic welding table plan. Some additions
you might want to include are roller wheels for each of the legs. Wheels
make the table more versatile.

You might also want to add more angle iron midway between the top
of the table and the floor or ground. You can then add a shelf to hold
scrap metal or other materials. Keep in mind, however, the potential fire
danger of storing materials under a table used for cutting and welding.

Another addition might be a 1/4-inch-thick steel plate, 24 inches long
and 8 to 12 inches wide. Attach this plate to one end of the worktable and
you've increased its versatility by providing a solid and unbreakable work
surface.

You can also attach hooks on one end of the tabletop edge. These can
be used to hang various frequently used tools such as a hammer, clamps,

and spark lighter. The point is to make your worktable as useful as possible. I would suggest that you build the basic table and use it for a period of time. Then, after you have a clear picture of your work surface requirements, add personal touches to make your work easier and possibly more enjoyable (FIG. 15-3).

OUTDOOR WORKSHOP SITES

An outdoor workshop can be almost anywhere around your home provided there is adequate space. As a rule, you must have access to electricity for running a grinding wheel, a drill, and possibly other equipment. You will also find it much easier if you can set up your equipment and begin work quickly instead of making a half-dozen trips to the house or garage to move the required materials and tools out to the work area. In most cases, this means your outdoor work space will be close to your house or garage.

Carport

A carport can make an ideal outdoor workshop. You can leave your welding equipment and tanks in one spot without worrying that the elements are going to play havoc with them. You can also store scrap and unused metal out of the weather. Ventilation will be of little problem as most carports are open on at least two sides. Just make certain that flammable materials, such as paints or other materials, are not in the area. In some

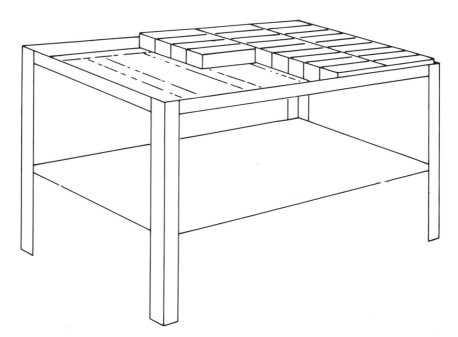

15-3 The finished welding table made from angle steel and steel strap.

cases, this means leaving your automobile and lawn mower in another location. If the carport has a concrete floor, be sure it is clean and free from oil or grease stains. If it has wooden walls, locate your welding table as far away from them as possible. You can cover existing walls with asbestos board, commonly sold in sheets from 2 to 4 feet wide and from 4 to 8 feet long.

Cement or stone patio

If you are not fortunate enough to have a covered carport, consider a cement or stone patio that is close to the house. Keep in mind, when using an uncovered area for welding projects, that the weather will determine when you can work. You must also rig up some means of covering the oxygen and acetylene cylinders if they are to be left outside (FIG. 15-4). In some regions, you can get by quite nicely by covering the equipment with a plastic tarp when it's not in use. In other parts of the country, this might not be the best protection. In the Great Basin (Western Deserts), for instance, rainfall is almost nonexistent, but in the Pacific Northwest inclement weather is quite common. You can also work on bare ground, if necessary, provided that all materials such as fallen leaves are removed from the area each time you use the outdoor shop.

Outbuilding

If you live on a farm or ranch, you will quite possibly have an outbuilding that can be used for a welding shop. You might also find that a number of welding tasks need to be accomplished in places other than the welding shop. In this case, you'll find it very handy to have your oxygen and acetylene tanks on some type of portable hand truck. You can then take the welding equipment where it is needed.

Whenever you work outside, especially on bare ground, you should clear the area of any materials that could catch on fire during welding or cutting operations. If you are working during the dry part of the year, it might be a good idea to wet the earth with a garden hose, before beginning work. Keep the hose close to the area you are working in so you will be able to quickly extinguish any fire that might occur. Needless to say, other fire-fighting equipment should be handy as well.

INDOOR WORKSHOPS

If outdoor space is not available or if you have suitable indoor space for welding, such as a garage, there are a number of things you need to consider before beginning any welding or cutting project. Adequate ventilation is an important prerequisite of any welding area. When you are working indoors, you must make certain that the air in the space is fresh and moving. In most cases, an exhaust fan of the type found in most kitchens will do the job quite nicely (FIG. 15-5). If you are working in a large area, you might find it necessary to install a heavy-duty, large-capacity fan to ventilate the room. The point is that all smoke and fumes must

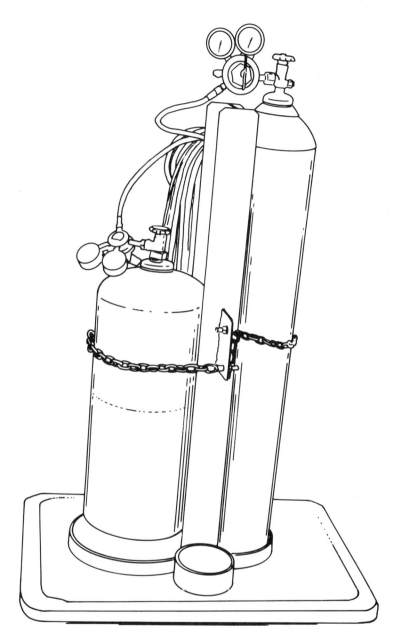

15-4 When oxygen and acetylene tanks are stored outdoors, they must be chained to a solid object, and covered when not in use.

be removed from the work area as quickly as possible (FIG. 15-6). Fresh air must come from outside of the room and not just be recirculated. You might, in addition to installing an exhaust fan, have to install a fresh air vent to introduce outside air and exhaust interior air.

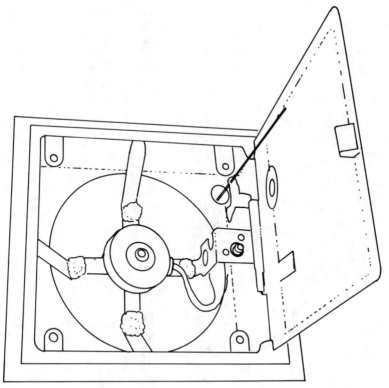

15-5 You must install an exhaust fan in an indoor welding shop to remove the fumes from welding and cutting operations.

Fire station

Your home workshop should have adequate fire-fighting equipment that is accessible and ready to use at all times. In most cases, it is best to designate one particular area as a ''fire station.'' Here, you should keep a fully charged fire extinguisher, a bucket of sand, and a pail of clean water. The fire station should never be used for anything like storage of scrap metal. There should always be a clear path to the fire-fighting equipment. If you are ever in need of your fire-fighting equipment, you will not only know exactly where it is but you will also be able to get to it as quickly as possible. This fire station might quite possibly save you from an out-of-control fire.

Accessibility

Still another requirement of an indoor welding workshop is that you must have accessibility for delivery of supplies, such as tanks of fuel gas and oxygen. If your welding shop is in a garage, there should be little problem in getting materials and supplies into the shop. If your welding shop is in

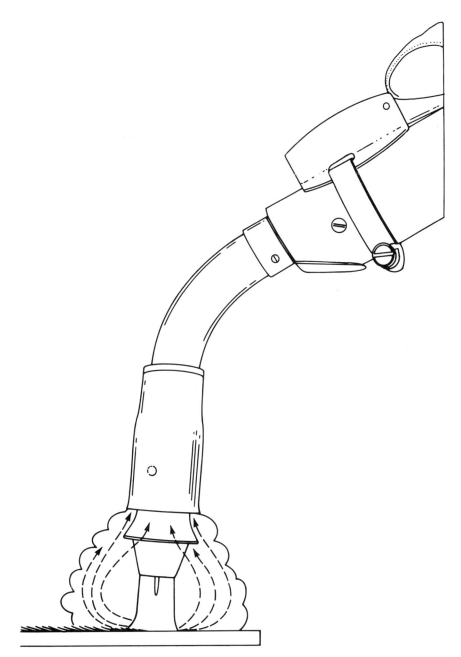

15-6 One manufacturer's solution to the smoke problem when welding is a smoke exhaust gun system. This system is used in industrial welding.

your basement, however, you might find deliveries a real exercise in equipment manipulation. Consider easy access as one very important prerequisite for any site you are planning to use as a welding shop.

Lighting

Lighting is important in any workshop, and a welding shop is certainly no exception. It is a fallacy to think you can work in a welding shop under a single light bulb. In addition to good lighting over your welding table, you must have direct light over your bench grinder and other work areas, such as where you solder or where your bench vise is located.

Probably the best type of lighting for a welding shop is fluorescent (FIG. 15-7). Two 4-foot-long tubes over the welding table should provide you with plenty of illumination. Fluorescent lighting is also the least expensive type of lighting to operate and run over time. For lighting of specialty areas, such as over a bench grinder, it is usually best to install a lighting fixture over each particular area (FIG. 15-8). Other lighting possibilities include a clamp-on reflector lamp or small fluorescent fixture.

If you find that you need a number of lighting fixtures, it will probably be best to install fluorescent lighting because this type of lighting runs cooler and is less expensive to operate in the long run.

While on the topic of electrical lighting, it might be a good time to mention that an indoor welding shop should have a number of electrical outlets. Tasks will be easier to accomplish if you can plug equipment into any one of several electrical outlets. Considering that there are a few electrical tools that might be commonly used around the welding shop, you will want to make provisions for them. You should have a few outlets in addition to those already in use by stationary equipment and machinery.

You might also want to install a 220-volt electrical outlet in the welding shop. Most arc welding units require this higher-line voltage for operation.

Storage space

Storage space is another important workshop consideration. It will become apparent as you spend more time welding that you will almost

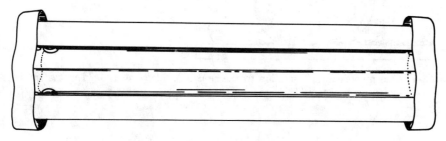

15-7 A fluorescent lighting fixture will provide good general lighting in the work area and will cost less than incandescent lighting.

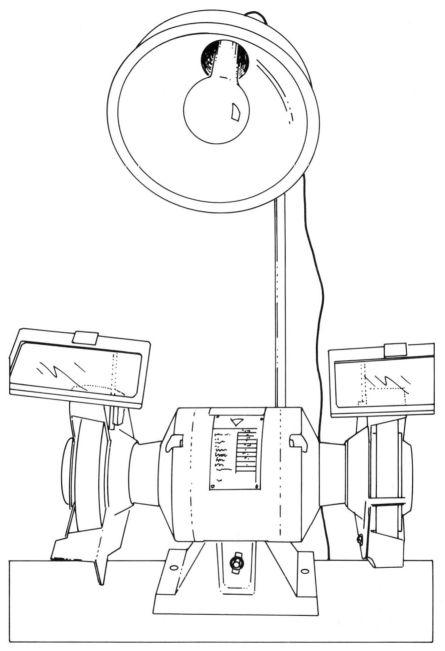

15-8 It is important to provide adequate lighting over your grinding bench. This clamp-on light is inexpensive and does the job.

always require more space than you have. Such is life. Just how much
space you really need will depend on your work habits and, even then,
the type of work you plan on doing. In all probability you will be limited
to a garage or other outbuilding plus the area outside the shop. Over a
period of time, you will undoubtedly accumulate metals, both new and
scrap, and you will want to store them out of the elements. You will also
probably find yourself with a number of tools and machines that are necessary for various metal-cleaning and joining tasks.

The obvious solution to "all that junk," as my wife is fond of saying,
is to build storage units in one area of the workshop. Sturdy shelves can
be used for storing small pieces of metal. Long pieces, such as angle iron
or strap, are best stored horizontal and flat on some type of rack system.
Plate and sheet metals can easily be stored standing on edge in some type
of vertical shelving system. Odd sizes and shapes of metal are usually a
problem to store. Often a metal garbage can, with three or four caster
wheels welded on the bottom, will take care of the problem. In all cases,
it is always best to store metal off the ground to prevent rust (FIG. 15-9).

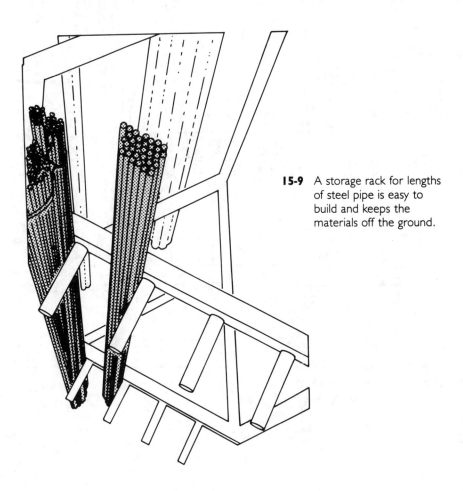

15-9 A storage rack for lengths
of steel pipe is easy to
build and keeps the
materials off the ground.

Very likely most of the metal you work with will be steel in one alloy or another. You might also have some nonferrous metals around the shop. It will be better in the long run to store all similar metals together. Some type of labeling system will be helpful in identifying these metals.

If your workshop is a small one, you will probably be forced to store much of your metal outdoors. To keep the metal from rusting, cover it with plastic or a canvas tarpaulin. Keep in mind that the best place to store metal is in some type of closed shed. The more serious you become about working with metals, the more likely you will build a storage shed.

ACCESSORIES

There are any number of useful projects that you can build around your welding shop. In addition to having specific areas for storing metals, you will need storage cabinets for tools and accessories. The next chapter covers in detail several woodworking projects of that kind.

Mobile cart

One project that I will briefly discuss right here is a storage unit for welding and brazing rods. Plans for a basic unit are given in FIG. 15-10. This unit can be made into a mobile cart by simply adding four caster wheels and a bottom panel to the basic unit. You will find such a cart quite handy for keeping your rods stored neatly as well as keeping them handy when you need them.

Anvil

Another useful accessory for your welding shop is some type of anvil. An anvil is quite handy for bending or flattening small work with a hammer. You can make one simply by cutting one end of a piece of scrap steel rail in the shape of a traditional anvil. A piece of rail about 16 inches long will be more than adequate for your needs. After the basic shape has been cut out of one end of the rail, pierce-cut four holes on the bottom with your cutting torch. Use these holes for fastening the anvil to a solid surface. One very good base for your home anvil is a section of tree about 18 to 24 inches in diameter and about 20 inches high. The log will absorb most of the hammering shock and help to deaden the noise. Finish off your anvil with a portable grinding wheel or disc sander.

LOCAL CODES

One last thing to consider about your home welding shop is the local codes. In some residential areas, a commercial welding shop is not permitted. It is in your best interests to check with your local authorities to see if any such restrictions exist in your area.

As a side note about restrictions in certain areas, you should know that if you call your shop a studio rather than a welding shop, you might find that the same restrictions do not apply. In many instances, your

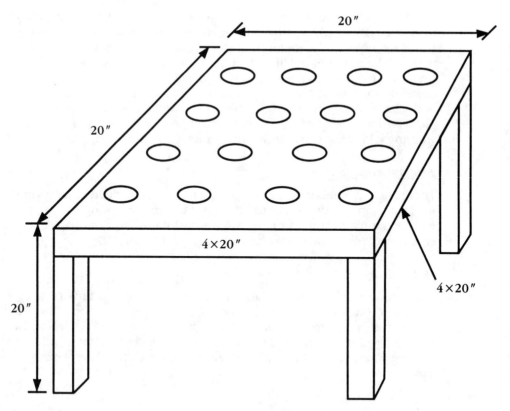

15-10   A welding and brazing rod holder table.

workshop will be a studio in the sense that it will be a place where you can express yourself through the art of metalworking.

SUMMARY

Choosing a location for your welding shop is a major consideration. If you limit yourself to small projects and only work occasionally, you will obviously be able to work outdoors. However, if you are planning to devote a fair share of your time to welding and cutting, you will need a space that can accommodate your needs. To decide if a particular area is suitable for general welding and cutting needs, consider the following eight requirements of a safe and useful welding workshop.

- **Ventilation.** Some means of exhausting smoke and gases must exist or be installed in the area. This exhaust system must have the capability of quickly clearing the air in the workshop as well as supplying adequate fresh air.

- **Fireproof construction.** The walls, floor, and ceiling of the welding shop must be covered with a noncombustible material to prevent fire.

- **Work area.** A good welding shop will have different areas for specific metalworking tasks. These might include a cutting and welding table, soldering workbench, metal grinding area, and possibly a separate area for finishing or painting metal.

- **Lighting.** Before you can expect to work in a workshop, you must provide adequate lighting.

- **Fire station.** Every welding workshop should have fire-fighting equipment in a central location.

- **Storage.** Welding, cutting, and related tasks such as grinding can be dusty affairs, to say the least. A good welding shop will have storage facilities not only for metals but for tools and equipment as well.

- **Access.** An effective welding shop will have wide doorways that permit easy movement of equipment and supplies.

- **Cleanliness.** A safe welding workshop is easy to keep clean and should be kept that way at all times. This will do a lot to prevent fire and also make for more pleasant working conditions.

In the final analysis, you will make the decision as to the best location for your own welding shop. Obviously, it will be to your advantage to choose a location that is adequate for your needs and enables you to practice the art of welding safely.

Chapter **16**

Metal projects

As you become more proficient at joining metals with the oxyacetylene process, you'll find an endless variety of metalworking projects to keep you busy. In this chapter, we'll look at 10 different metalworking projects you can build with your welding equipment.

PLANT STAND

Metal plant stands are easy to construct with modern welding equipment and provide a sturdy base for even the largest plants. Metal stands can be used indoors or outdoors and can be painted any color to match or contrast with the surroundings (FIG. 16-1).

Keep in mind that the dimensions given for this plant stand project are not etched in stone. If you have a specific pot in mind for this planter, make the top hoop a corresponding diameter. The plants offered are for a standard 9-inch-diameter clay pot—which has a top diameter of 8½ inches. The inside diameter of the hoop given in the illustration should match this for a snug fit. If you plan on using a larger or smaller clay or plastic pot, adjust the dimensions of the top hoop accordingly. The top hoop, of course, holds the pot and prevents it from falling through.

To make the plant stand, you will need approximately 15 feet of ⅛-inch-thick × ½-inch-wide steel strap. Cut three pieces 38 inches long (for the three legs) and one 10-inch piece, one 22-inch, and one 26-inch for the hoops.

Begin by bending the 26-inch length of steel strap into a circle shape. Once you are satisfied with the shape, clamp the ends together; then tack-weld. This is the top hoop and, as mentioned, it should be wide enough to hold a flower pot securely. Next, form circles from the other two pieces (22-inch and 10-inch) and tack-weld.

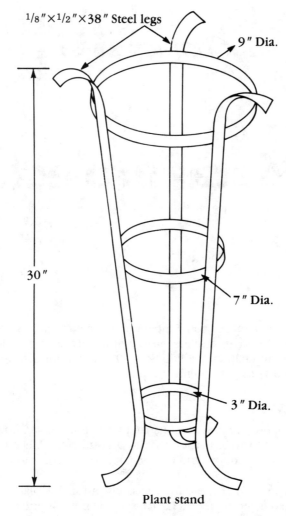

1/8″×1/2″×38″ Steel legs

9″ Dia.

30″

7″ Dia.

3″ Dia.

Plant stand

16-1   Plant stand.

The legs

Each of the 38-inch-long pieces of steel strap must be heated approximately 4 inches up from the ends and bent to form a foot. Similar treatment, as you can see from the illustration, is given to the top of each leg. Completed legs will be 30 inches long.

Curving the ends of the legs is not hard. Apply heat evenly until the metal turns red-hot. Use pliers to make the bend. If you're not satisfied with the shape of the curve, wait until the metal cools, then bend the strap until you achieve an adequate curve. Once you have bent one leg—both top and bottom—duplicate the bends on the other two legs.

After the three legs are bent to your satisfaction, they can be attached to the hoops. Begin by positioning the legs at three equidistant spots around the outside surface of the top hoop. Clamp in place, then spot-weld. Next, position the center hoop, and spot-weld. Finally, give the same treatment to the bottom hoop.

To finish, grind off any excess welding rod with a wire brush. Prime and apply two coats of black enamel—or any other color you choose.

WINDOW SECURITY BARS

Residential theft has been increasing for many years, and access is commonly made through windows. While electronic security systems are common in many homes, these systems do not block entry, and clever criminals can often override them. Window security bars offer a substantial barrier and can be constructed and installed for a fraction of the cost of an electronic security system. When both systems are used, your home should be almost totally safe from intruders.

While window bars tend to obstruct the view somewhat, many homeowners feel this slight inconvenience is worthwhile, especially in high-crime areas. As a rule, window bars are used only on basement and ground-floor windows, and therefore, the view from second-story windows is not a problem. Window security bars are rarely used on casement windows (those that open outward) because the bars prevent normal operation. Single- and double-hung windows are best suited for security bars because they can be opened in an up and down manner. The same is true of fixed glass windows (FIG. 16-2).

Window security bars must be custom-made for each ground-floor window in the home, but the basic design is the same for all windows. One-inch steel channel is used for the top and bottom rails and $3/4$ or $7/8$-inch steel bars are used for the upright pieces. The uprights are spaced about five inches apart and should be perpendicular to the top and bottom channel pieces.

For windows up to 3 feet in height, the security bars are simply constructed from top and bottom channels and upright bars. For windows with a height greater than 3 feet, it is common to add a horizontal bar about midway between the top and bottom channel to add strength. A suitable length of steel bar can be used for this. A horizontal bar also prevents a thief from pulling the bars apart to gain entry.

Constructing the security frame

To construct each window security frame, measure the sill and header of the window. These pieces should fit fairly snugly with no more than about $1/2$ inch of space between the ends of the channel and the window frame. Next, measure and cut the upright bars. Remember that these will be fitted into the channel; so make allowances accordingly—it might be necessary to grind the tips of the bars, for example, so they will fit into the channel.

Lay all the parts on a flat surface and fit the bars into the channel with a space of about 5 inches between members. Clamp the bars into the

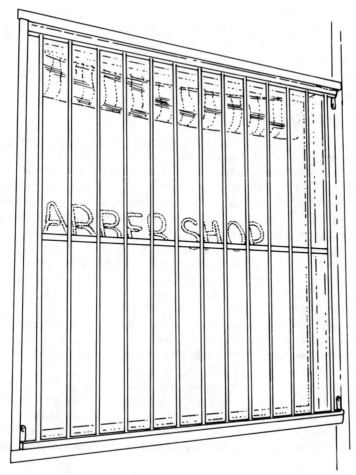

16-2   Window security bars over a fixed glass window.

channel; then tack-weld the bars in place. Check your alignment often to make certain that the upright bars are perpendicular to the top and bottom channel pieces. A framing square is handy for this task.

Assembly and finishing

Once the frame is constructed and cooled, check the fit in the window frame. Next, predrill holes for fastening to the top header and bottom sill plate. Before installing, prime and give two coats of metal paint. If your windows are white, you might want to paint the frames white as well.

Once the paint is dry, you can install the security bars into the window frame. Fasten the top channel into the header above the window and the bottom channel into the sill of the window. This should be done through the predrilled holes using 4-inch-long lag bolts.

METAL RAILING

A metal railing on your front or back steps adds safety, and a decorative touch to the entryway. A metal railing can also be used around wooden decks for the same reasons and will offer security without obstructing the view as much as a wooden railing.

A basic metal railing (FIG. 16-3) is composed of a top and bottom rail connected by upright steel bars. A more decorative railing could easily be constructed by adding curved steel rod or strap intermittently—these are bent into the desired shape and attached between the upright bars or right over the bars.

There are two basic types of steel railings: straight and angled. Straight railings, as the name suggests, are used on porches and decks.

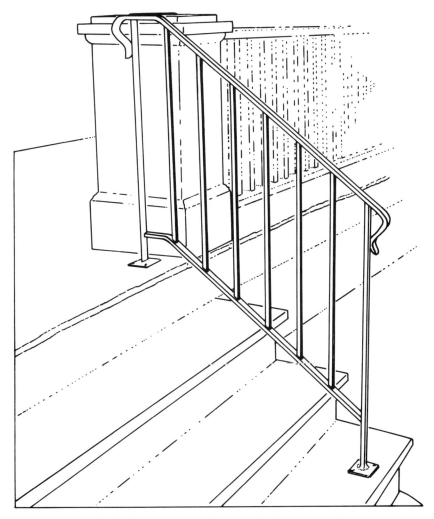

16-3 A basic metal railing for stairs is easy to weld.

The top and bottom rails are parallel and the upright bars are fastened perpendicular to these. Angled railings also have parallel top and bottom rails but because they are installed on a sloping surface—steps, for example—the bars are vertical but not perpendicular to the top and bottom rails. It is common to have both types of railings in one installation, such as where a railing comes up steps and joins a porch railing.

The steel used for metal railings should be easy to obtain from your local steel fabricator. In some areas, you might want to shop for components at a railing building shop, where you can usually find decorative top rails as well as steel decorations. The top rail should be decorative on the top surface and have a channel on the underside. The bottom rail is most commonly made from steel channel, 3/4 to 1 inch wide and 1 inch deep. Square steel bars (3/4 to 7/8 inch) are used for the upright members.

The standard measurements for a steel railing, as shown in FIG. 16-4, call for a finished height of 36 inches. The upright bars will measure 32 inches and the fastening bars must be at least 40 inches if encased in cement. If the fastening bars are attached to plates, they should be 36 inches long. Bar spacing should be about 5 inches. Exposed ends of top railings should be longer than required. This will allow you to heat the area and bend the excess top rail for decoration.

Metal railing

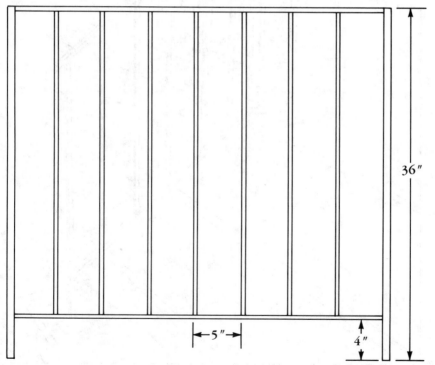

16-4 Dimensions for basic, straight railing.

To construct a straight steel railing, begin by determining the length of the top and bottom rails. Next, cut and lay out the pieces on a flat surface. Remember that the top and bottom rails are parallel and are spaced 32 inches apart. Upright bars are also 32 inches long and are spaced about 5 inches apart, perpendicular to the top and bottom rails. On both ends, and spaced about 4 feet apart, you must install longer upright bars for attachment to the deck or concrete. These longer bars replace the standard bars and must pass through the bottom rail.

Once the layout of the pieces is to your satisfaction, mark the location of the longer fastening bars on the bottom rail—on both ends and about 4 feet apart along the railing. Since these bars must pass through and extend below the bottom rail, suitable-size holes must be made at these locations. This is most easily accomplished using the plunge or piercing cut with your cutting torch. Make these holes slightly larger than the size of the steel bars, and fill in later with welding rod.

Lay out the parts again, and begin by attaching the longer fastening bars through the bottom rail and under the top rail. Hold the pieces in place with clamps and simply tack-weld in place. Check alignment often with a framing square to make certain that the bars are attached perpendicular to both the top and bottom rails.

Attaching a straight railing can be done in a number of ways, depending on the type of deck or porch. For new concrete, simply set the railing in place before the concrete hardens. You should also attach the end or ends to the building with lag bolts through predrilled holes (FIG. 16-5).

To install a railing on old or hardened concrete, you must first predrill holes in the deck or porch. These holes should be fairly close to the size of the attachment bars. In setting the railing, first fill these holes with a concrete patching mixture; then insert the attachment bars into the holes. When the patching material hardens, it will securely hold the railing in place. You should also attach one or both ends of the railing to the building.

One common method used to attach a metal railing involves special decorative feet or plates. These are available from railing building shops and specialty catalogs. For this method, the attachment bars extend only four inches below the bottom rail and are inserted into the pre-installed plates. A set screw through the plate holds the attachment bar in place. The ends of the railing should also be attached to the building with lag bolts.

You can also attach small steel plates (3/8 inch thick, 3"×3") to the feet of the railing for securing to concrete or wooden decks. Predrill the holes and lag bolt in place (FIG. 16-6).

Angled railings (FIG. 16-7), which are common on stairs, are very much the same as straight railings with the exception of the angle of the upright bars. In almost all cases, the bars are attached to the top and bottom rails at a 45-degree angle. This means both the bottom and top end of the steel bars must be ground to approximately this angle before they are fastened to the rails. It is also important to keep these bars at a vertical angle and spaced about 5 inches apart.

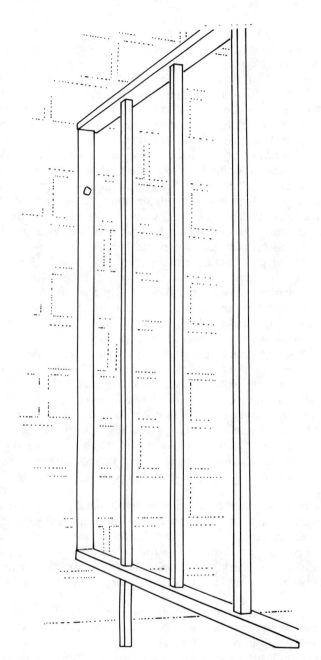

16-5   Attach the end of the railing to the building with lag bolts.

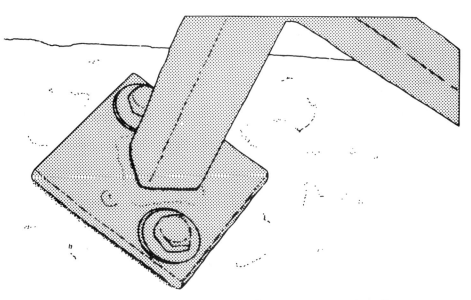

16-6 Steel plate attached to legs of railing allows you to attach the railing to concrete or wooden decking.

Angled railings are attached by any of the means covered earlier: in new concrete, or else drilled in old concrete or into plates. Where an angled railing joins a straight railing, the joint is commonly welded. But where the angled railing simply joins a building, the best method of securing is to lag-bolt the end of the railing to the building.

After the railing has been constructed, use a wire brush or disc sander to clean up the welds and remove any rough edges. In areas that receive significant amounts of rainfall, you should drill holes in the bottom channel—between the upright bars—to allow drainage of rainwater. Next, give the railing a coat of metal primer followed by two coats of exterior metal paint. The standard color for metal railings is black enamel. All priming and painting should be accomplished before the railing is installed.

BICYCLE RACK

A steel bike rack should last a lifetime and provide an excellent means of storing bikes when not in use. In addition, your bikes will be much more secure if locked to a steel rack rather than allowed simply to sit outdoors or in the garage.

The plans in this section (FIG. 16-8) call for a triangular-shaped, free-standing bicycle rack, which is best suited for use against a wall in a garage, outbuilding, or outdoors. While the basic plan calls for a rack that will hold four bikes, it can be expanded. If your storage needs are greater, simply increase the overall length of the rack by adding 8 inches for each bike. This rack will hold any size bicycle securely.

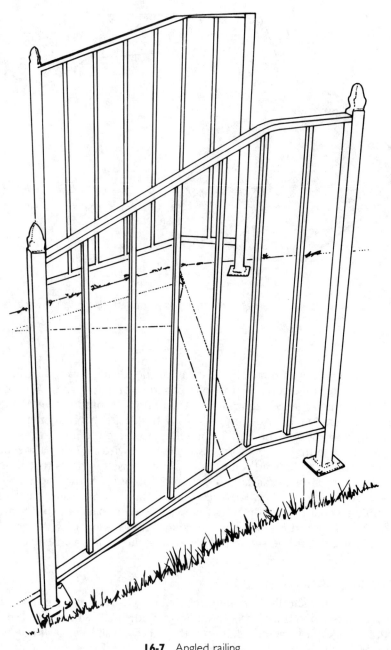

16-7   Angled railing.

Bicycle rack

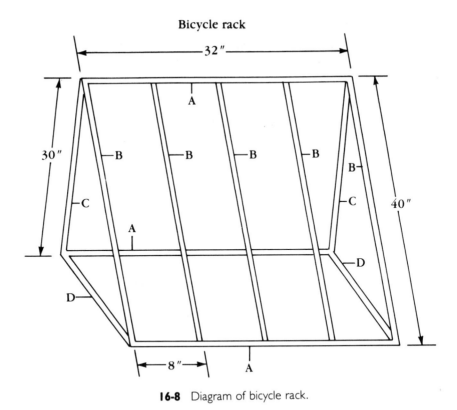

16-8 Diagram of bicycle rack.

To build this bicycle rack you will need 34 feet of ¹/₂-inch steel bar. Cut the pieces as follows:

MATERIALS LIST

32″—top and two bottom rails	(3)
40″—angled support bars	(5)
30″—rear uprights	(3)
26″—side supports	(2)

Begin assembly by attaching the side supports (d) to the rear uprights (c) being careful to join these at a right angle. Next, attach the top and two bottom rails to the rear uprights for the basic frame of the rack. Lastly, attach the angled support bars to the top and bottom rails, allowing for a space of 8 inches between these bars.

After the bicycle rack has been constructed, use a wire brush or disc sander to clean up the welds and remove any rough edges. Then apply a coat of metal primer, followed by two coats of exterior metal paint. All priming and painting should be accomplished before the bicycle rack is pressed into service.

METAL SAWHORSES

Since so many welding projects are easier to accomplish at waist level than on the ground, it will be to your advantage to build a pair of sturdy metal sawhorses for the workshop. The plans given in this section (FIG. 16-9) will result in a pair of durable sawhorses that will enable you to place even large projects at a comfortable working height.

To make a pair of the illustrated metal sawhorses you will need the following materials:

MATERIALS LIST

a) 48"   2"-×-2" square steel tubing, top rail (2)
b) 30"   2"-×-2" square steel tubing, legs   (8)
c) 16"   2" steel strap, braces   (4)

Begin by cutting the ends of all legs (8) at a 30-degree bevel. First mark the cutlines using a protractor. Then make the cuts using a hacksaw. Clean up the cut edges with a bench grinder or disc sander to remove any metal burrs. The fit to the top rail of the sawhorse should be two well-mated surfaces. Once you have accomplished this you can begin welding the legs in place.

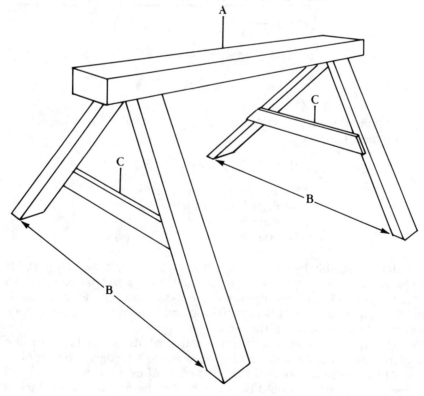

16-9 Metal sawhorse.

Position one leg (b) about 6 inches in from the end of the sawhorse top rail (a). Use a clamp to hold the leg in this position and tack-weld the top of the leg to the top rail. The top of the leg must be flush with the top rail. Once you have done this you can remove the clamp and finish welding. Run a welding bead around the entire joint, using filler rod for a string weld.

After the first leg has been attached in this manner, fasten another leg on the opposite side of the top rail. Repeat this procedure on the other end of the top rail, welding those two legs about 6 inches in from the end. Once all four legs have been attached, stand the sawhorse up and inspect the work. It should stand firmly, if all welds have been made properly.

The next step is to attach the horizontal braces between both pairs of legs. Place the steel strap (c) over the inside edges of one pair of legs and adjust so it is horizontal to the ground. Hold the brace in place with clamps while you tack-weld in position. Remove the clamps and run a welding bead around all mating surfaces of brace and leg. Repeat this procedure for the opposite pair of legs.

After you have welded one sawhorse, weld the second horse in the same manner. You will then have a pair of metal sawhorses that should last a lifetime.

Before you press your new sawhorses into service you should clean up all the welds and any sharp edges with a wire brush and disc sander. Follow this with a coat of metal primer and two coats of metal paint, if desired. This will protect the metal from rust and give a nice finish appearance.

METAL GATE

A metal gate is particularly handy around the farm or ranch but can also be useful in a residential setting—at the end of the driveway or as a gate for the backyard fence. The diagram (FIG. 16-10) in this section is for an

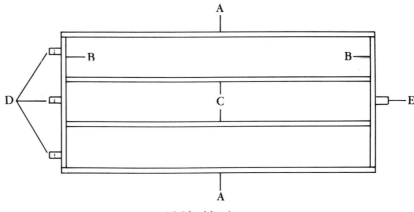

16-10 Metal gate.

8-foot long gate but can, of course, be modified for your particular needs simply by adjusting the lengths of uprights and cross members.

This gate is attached to an upright pole with standard gate hardware. The female parts are welded on one of the side uprights, and the lag screw male parts are placed into the pole at suitable locations. For long gates, such as in the illustration, three attachment points are recommended. For smaller gates—up to about 6 feet in length—two attachment points should suffice. The latch can also be standard gate hardware or, if security is a consideration, a chain and lock.

The pole that holds the gate up must be painted deeply, or the gate will sag in time. A hole with a minimum depth of about 6 feet is required for this gate, with the pole set in concrete as well.

MATERIALS LIST

a) $60''$-×-$1^1/2''$-×-$1^1/2''$ square steel tube, sides   (2)

b) $8'$-×-$1^1/2''$-×-$1^1/2''$ square steel tube, top and bottom rails (2)

c) $7'1/2''$-×-$1^1/2''$-×-$1^1/2''$ square steel tube, middle rails   (2)

d) Lag bolt gate hinges   (3)

e) Gate latch or lock and chain   (1)

Begin by cutting (a) top and bottom rails (b) and middle rails (c) to the proper length with a hacksaw. After cutting, clean up all cut edges with a disc sander so that all mating surfaces are beveled slightly and form a tight-fitting butt joint. Next, lay the parts out on a flat surface—the two longer rails are placed on the top and bottom of the ends, and the shorter rails are positioned at equal distances apart in the center of the gate. It is important that the gate be square, so use a metal framing to help you align the pieces.

Begin by tack-welding the top and bottom rails (b) over the ends of the uprights (a). Check the alignment often with your framing square to be sure the basic outline of the gate remains square. After all parts have been tacked, weld each joint solidly by running a bead of welding rod all the way around these butt joints.

Next, reposition the middle rails (c) in the gate. Make certain that these pieces are parallel to the top and bottom rails. Then tack-weld in position. The next step is to weld the middle rails in position by running a continuous bead of welding rod around each of the butt joints.

Attach the female gate hinge parts by first drilling holes at suitable locations. Insert the hinge parts and weld in place. If a latch is to be attached to the other upright, do this next.

After the gate has been welded, clean up all joints and edges with a disc sander so there are no sharp edges. Next, give a coat of metal primer to all surfaces followed by two coats of exterior metal paint. You might find the priming and painting easier to accomplish if you install the gate beforehand.

WINE RACK

If you enjoy drinking fine wines—as well as making them yourself—you will want a sturdy and secure place to store bottles of wine in your cellar. The plan in this section (FIG. 16-11) results in a metal wine rack that will hold approximately 99 standard wine bottles. The rack is freestanding and holds each bottle in a horizontal position, which experts claim is the proper position for storing corked wines.

This wine rack is made entirely from 1/2"×1/2" angle steel. A total of 90 feet of this material is required for the project. One-half-inch angle steel is readily available from any metal fabricator and many lumberyards. This soft steel is easy to weld, and that makes this project a good one for the beginning welder.

While the wine rack project offered in this section is designed to be freestanding, it is wise to fasten the two tops of the rear legs to a wall. This will prevent the rack from ever falling over. The rack itself is utilitarian in nature, with no decoration. If your wine cellar is large, make more than one unit instead of trying to expand on the basic plan.

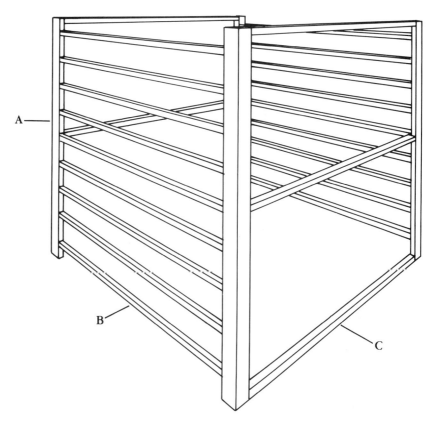

16-11 Metal wine rack.

MATERIALS LIST
(all parts from 1/2"-×-1/2" angle steel, 90' required)

a) 72" legs   (4)
b) 36" racks   (18)
c) 8" top, bottom, and middle braces   (6)

Begin by cutting all the pieces with a hacksaw. Clean up all cut ends with a bench grinder so there are no sharp metal edges. Next, lay two legs (a) parallel on a flat surface with the top and bottom ends aligned. They should be 36 inches apart. Then lay the rack pieces (b) in place. The first one is placed 3 inches up from the bottom, and the remaining pieces (8) are placed at an 8-inch spacing. Each piece is positioned so the flat surface is on top—this is where the wine bottles will rest.

Tack-weld each rack in two spots at each joint. This should be sufficient for holding them in place and generally reduces the amount of heat required for the project. Excessive heat could have a tendency to warp the upright legs.

After the front of the rack has been completed—nine pieces tack-welded in place—turn your attention to the rear of the rack. Each of these pieces is tack-welded in position at the same spacing, but the flat edge of the angle iron must be down. This will prevent the wine bottles from falling, as the flat edge will provide a notched shelf for the bottles.

Once the front and back have been tack-welded, attach the top, bottom, and middle braces (c). The middle braces should be located approximately midway up the sides, or about 36 inches up from the bottom.

After the wine rack has been assembled, clean up all edges, joints, and welds with a disc sander. Apply a coat of metal primer followed by two coats of metal paint. When you install the wine rack in a suitably cool basement, drill two holes in the tops of the rear legs and lag-bolt the rack to the wall. You are now ready to store 99 bottles of wine.

AUTOMOTIVE ENGINE HOIST

Major automotive engine work commonly entails removing the engine from the vehicle. While this can be done in a number of ways, the work will go much quicker and be safer to accomplish if you use a specially-designed engine hoist. To be sure, these are available from rental companies, but if you do much of this type of work, it will be to your advantage to custom-make an engine hoist for your workshop. The materials are readily available, and the project is not difficult for those with some welding experience. This is not a good welding project for the beginning welder, however, as the thicker metals used are challenging to weld (FIG. 16-12).

MATERIALS LIST

a)   55"-×-2"-×-3" square steel tube—boom   (1)
b)   60"-×-3"-×-3" square steel tube—upright   (1)

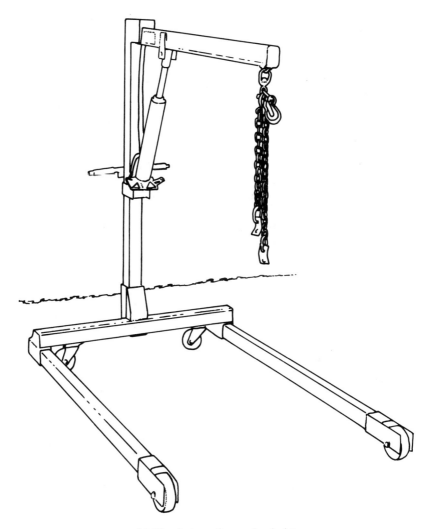

16-12 Automotive engine hoist.

c) 40"-×-3"-×3" square steel tube—rear base (1)
d) 65"-×-2"-×-3" square steel tube—legs (2)
e) 8"-×-3½"-×-2½" square steel tube—connectors (2)
f) 5"-×-4"-×-³/₈" steel plate—boom braces (2)
g) 3"-×-½" steel rod—boom connector (1)
h) 10"-×-3¼"-×-3¼" upright connector (1)
i) 6"-×-4" steel channel—brace for upright (1)
j) 9½"-×-3"-×-³/₈" steel plate—front wheel brackets (4)
k) 5" steel wheels—front wheels (2)
l) 4" steel wheels with swivel brackets—rear wheels (2)
m) 17"-×-³/₄" steel pipe—handle (1)
n) hydraulic jack—16" lift (1)

o) 5"-×-2¹/₂"-×-³/₈" steel plate—jack connector   (2)
p) 2¹/₂"-×-1¹/₂" steel tube—jack connecting point   (1)
q) 2" steel rod for attaching (p) above inside plates   (1)
r) 7"-×-2"-×-¹/₂" steel plate, U-shape—boom connect   (1)
s) 6"-×-¹/₂" steel rod—bent to form eyelet for chain   (1)
t) 36" steel chain and hook   (1)
u) 5"-×-3"-×-³/₈" steel plate—jack connecting plates   (2)
v) 3"-×-¹/₂" steel rod—jack connecting shaft   (1)

Gather together all of the listed materials, cutting and cleaning up cut edges, to remove burrs and rough edges, where required. Begin welding the base of the hoist by attaching the square tube connectors (e) to both ends of the rear base (c) of the unit. Next, in the top middle of the rear base, attach the upright connector (h). Attach the brace (i) in front of the upright connector. Insert the 60" square tube—upright (b)—into the upright connector and weld in place. At the top of the upright, attach the boom braces (f) on both sides. These should be flush with the top and protrude forward. Midway between these plates, weld in the steel rod boom connector (g). Lastly, attach the handle (m), midway up the upright (FIG. 16-13).

Begin assembling the two legs (d) by attaching the front wheel brackets (j) to both sides of each leg. Drill holes for the 5" steel wheels (k), and install these wheels using suitable-size bolts as axles. Now insert each of the legs into the rear frame brackets and weld in place. Next, attach the rear wheels (1) to the underside rear base, 5 inches in from each end.

Just above the handle on the upright, attach the jack connecting brackets (u) to both sides. Then between these, weld in the jack connecting shaft (v). This should be located approximately 2 inches from the end and centered between the plates. The hydraulic jack (n) should fit on top of this shaft.

Now turn your attention to the bottom of the hoist. Begin by attaching the boom connect (r) on top of one end of the boom (a). Next, about 18 inches from this end of the boom attach the two steel plate connectors (o). Between these plates install the steel rod (q) and the steel rod (p) that is the female connector for the top of the hydraulic jack. Bend the steel rod(s) into an eyelet shape and weld in place on the lower side of the boom, about 4 inches from the far end. Lastly, attach the chain and hook (t) to the eyelet (FIG. 16-14).

After all welding has been completed, clean up all welds with a disc sander to remove any roughness. Prime the hoist with metal primer, followed by two coats of metal paint. The engine hoist is ready for service and should provide years of dependable service.

ENGINE STAND

It is far easier to work on an automotive engine when it is mounted on a special stand than while it is simply on a workbench or on the ground. The engine stand offered in this section is mounted on swivel wheels that

**Boom
detail**

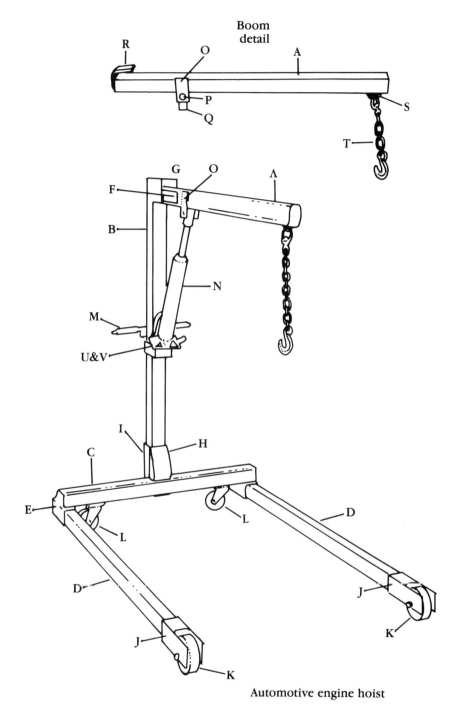

Automotive engine hoist

16-13 Diagram of automotive engine hoist.

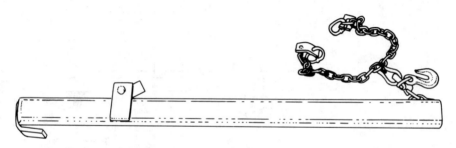

16-14 Boom for automotive engine hoist.

allow you to move an engine around the workshop. In fact, once an engine has been removed with a hoist, it can then be bolted, through the back of the engine block, to this stand and moved about quite easily (FIG. 16-15).

MATERIALS LIST

a) 37"-×-2$^{1}/_{2}$"-×-2$^{1}/_{2}$" square steel tube—rear base   (1)
b) 6"-×-2$^{1}/_{2}$"-×2$^{1}/_{2}$" square steel tube—top mount bracket   (1)
c) 6"-×-2$^{1}/_{2}$"-×-2$^{1}/_{2}$" square steel tube—front mount bracket (1)
d) 34"-×-2"-×-2" square steel tube—front leg   (1)
e) 28"-×-2"-×-2" square steel tube—upright   (1)
f) 4$^{1}/_{2}$"-×-2$^{3}/_{4}$" steel tube—top mounting bracket   (1)
g) 4" steel wheels on swivel brackets   (3)
h) 10"-×-6"-×-$^{1}/_{2}$" steel plate—mounting plate   (1)
i) 5"-×-2$^{1}/_{4}$" steel tube—swivel bracket   (1)
j) 1"-×-1"-×-5" square steel rod—engine mounting bracket   (4)
k) 1"-×-3" hollow steel rod—engine mounting bracket shaft   (4)

Begin by cutting all the pieces with a hacksaw. Clean up all cut ends with a bench grinder so there are no sharp metal edges. First weld the front mount bracket (c) in the middle of the rear base (a). Next, weld the top mount bracket (b) in place on top of the rear base (a). Then insert the front leg (d) into the front mount bracket (c) and weld in place (FIG. 16-16).

Turn the base of the engine stand over and attach the wheels (g) 3 inches in from the outside of each end. The wheel bases can simply be spot- or tack-welded in place, but be careful not to warp the thin metal of the base. As an alternative, the wheel bases can each be screwed into predrilled holes with no danger of warping the base metal.

Place the engine stand right side up and insert the upright (e) into the top mounting bracket (b). Weld the upright in place at a right angle to the base. Next, place the top mounting bracket (f) on top of the upright and align so that the back edge is flush with the back of the upright and the steel tube protrudes forward. Weld the top mounting bracket in place.

After the base and upright of the engine stand have been welded, you can turn your attention to constructing the mounting plate. Begin by mak-

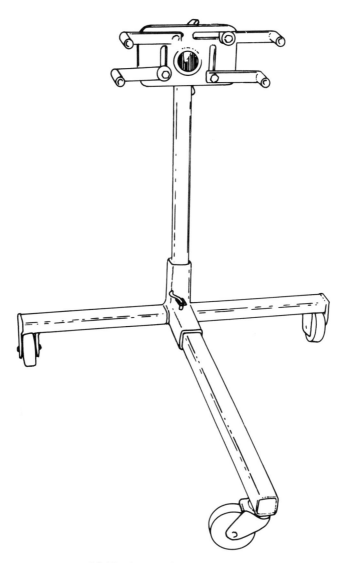

16-15 Automotive engine stand.

ing a pierce cut in the center of the steel mounting plate (h) approximately
2¼″ in diameter. Then insert the swivel bracket (i) into this hole and weld
with the back edge of the tube flush with the face of the mounting plate.

Next, you must weld the four engine mounting brackets (FIG. 16-17).
This is accomplished by simply welding an engine mounting bracket shaft
(k) to the end of each of the engine mounting brackets (j). After all the
mounting brackets have so been constructed, each must have a ¾″-inch
hole drilled through the face. This hole is located about 1 inch in from the
end of the square rod.

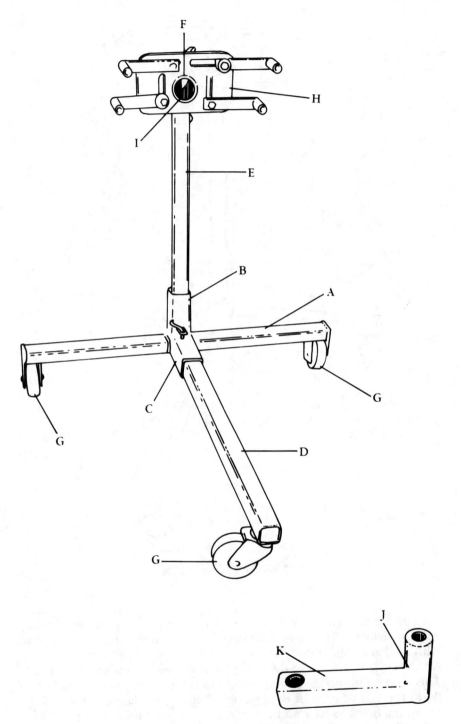

16-16   Diagram of automotive engine stand.

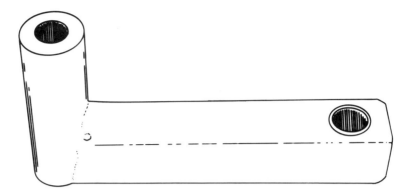

16-17 Engine mounting brackets—four are required.

Holes must be drilled in the mounting plate to accommodate the engine mounting brackets. These should be ¾ inch in diameter and located approximately 2 inches in from each corner. Then attach the engine mounting brackets through these holes.

Insert the engine mounting plate swivel bracket (with engine mounting brackets in place) into the top mounting bracket on top of the engine stand. Adjust so the top of the plate is level, and mark this location. Then drill a ⅜-inch hole through the top of both brackets. These holes permit the use of a metal pin that will hold the bracket in this position while you are working on an engine. Next, turn the plate 90 degrees and drill another hole through both brackets. This hole will permit the engine to be rotated and locked in place with a pin during work.

After all welding has been completed, clean up all welds with a disc sander to remove any roughness. Prime the engine stand with metal primer, followed by two coats of metal paint. Your engine stand is ready for use and should provide years of dependable service.

METAL WORKBENCH

Any workshop needs a sturdy workbench for working on projects. A workbench for an automotive or metal welding shop has to be extremely heavy-duty as the work you will be doing here commonly involves material of substantial weight. The metal workbench offered in this section will provide a rugged work surface that is unaffected by hard work and should provide a lifetime of sturdy service. Additionally, this is a good beginner welding project (FIG. 16-18).

MATERIALS LIST

a) 28″-×-2″-×-2″ steel angle—legs (4)
b) 48″-×-2″-×-2″ steel angle—front and rear top supports (2)
c) 24″-×-2″-×-2″ steel angle—side top supports (2)
d) 48″-×-1½″-×-1½″ steel strap—front and rear rails (2)
e) 24″-×-1½″-×-1½″ steel strap—side rails (2)
f) 4′ 4″-×-2′ 4″-×-⅜″ steel plate—top (1)

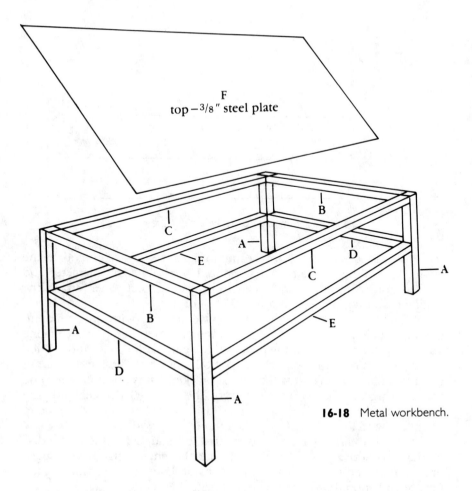

F
top – 3/8 " steel plate

16-18   Metal workbench.

Begin by cutting all the pieces with a hacksaw. Clean up all cut ends with a bench grinder so there are no sharp metal edges. When you purchase the steel plate for the top of this workbench, have it cut and save yourself a lot of work. If this is not possible, cut the top after it is mounted on the frame.

Start welding the frame by attaching the front supports (c) to the tops of two legs (a). Repeat this for the rear legs and rear top support. Next, attach the top side supports (c) between the front and rear sections. You will find all of this work easier to accomplish if you clamp the pieces together before welding. Run a continuous bead of welding rod around the edges of the mating surfaces.

After the top and legs of the frame have been welded, attach the front rails (d) about 18 inches up from the bottom of the front legs. Install the other long rail between the rear legs in the same manner. Then attach the side support rails (e) between the side legs.

Once the basic frame has been constructed, mount the top (f). All edges should protrude 2 inches over the top supports, all the way around the workbench. Tack or spot-weld the top in place. Place these small welds about 6 inches apart on the underside—where the top sits on the top supports. Be careful not to use too much heat as this could cause the top sheet steel to warp.

After all welding has been completed, clean up all welds with a disc sander to remove any roughness. Sand the top smooth to remove any rust or rough spots as well. Prime the workbench with metal primer, followed by two coats of metal paint. After paint dries hard, this workbench is ready for use and should provide years of sturdy, dependable service.

Glossary

abrasion A condition of wear most often caused by rubbing together two or more surfaces.

ac The accepted abbreviation for *alternating current*. A common type of electricity that reverses its direction of electron flow regularly and periodically.

acetylene Gas composed of two parts carbon and two parts hydrogen. When burned in an atmosphere of oxygen, it produces one of the highest flame temperatures.

acetylene cylinder A specially built container used to store and ship acetylene. Occasionally called *tank* or *bottle*.

acetylene regulator An automatic valve used to reduce acetylene cylinder pressures to working torch pressures and to keep the flow of acetylene constant.

acid core solder See *cored solder*.

activated rosin flux A rosin or resin base flux that contains an additive to increase wetting action of solder.

alloy A mixture of two or more metals to achieve specific qualities, such as hardness, ductility, and so forth.

annealing Softening metals by heat treatment. This process most commonly involves heating the metals up to a critical temperature and then cooling them slowly.

anode The positive terminal of an electrical circuit.

arc The flow of electricity through a gaseous space or air gap.

arc cutting A group of cutting processes wherein the severing or removing of metals is accomplished by melting with the heat from an arc between an electrode and the parent metal.

arc-seam weld A weld bead with an arc welding unit.

arc-spot weld A spot weld made by an arc welding process.

arc voltage The electrical potential across an arc. The pressure or voltage of an arc.

arc welding   Fusing metals using the arc welding process.

automatic oxygen cutting   Oxygen cutting with equipment that is fully automated, requiring only that an operator set up the work initially.

automatic welding   Usually, some type of arc welding wherein all welding operations are controlled and initiated by automation.

AWS   The abbreviation for the American Welding Society.

axis of a weld   An imaginary line along the center of gravity of the weld metal and perpendicular to a cross section of the weld metal.

back gouging   The forming of a bevel or groove on the other side of a partially welded joint to assure penetration upon subsequent welding from that side.

backfire   Momentary retrogression or burning back of the torch flame into the torch tip. Immediately following the withdrawal of the tip from the work, the gases can be reignited by the hot workpiece. Otherwise, the lighter might be necessary.

backhand welding   That method of welding in which the torch and rod are so disposed in the vee that the torch flame points back at the completed weld, enveloping the newly deposited metal. The rod is interposed between the torch and the weld.

balling up   The formation of globules of molten brazing filler material or flux caused by failure to adequately wet or tin the base metal. Also a professional welders' term (slang) used to describe a job that has been done poorly.

base metal   Materials composing the pieces to be joined by welding. Also called *parent metal*.

bead   Denotes the appearance of the finished weld and describes the neatness of the ripples formed by the metal while it was in a semiliquid state.

bevel   A special preparation of metal that is to be welded; here the edge is ground or cut to an angle other than 90 degrees to the surface of the parent metal.

blind joint   A joint in which no portion is visible.

blowpipe   A term used to describe an oxyacetylene torch handle.

bond   Junction of the weld metal and the base metal.

braze   A weld wherein coalescence is produced by heating to temperatures higher than 800 degrees Fahrenheit and by using a nonferrous filler metal with a melting point below that of the base metals. The filler metal is distributed in the joint by capillary attraction.

braze-welding   A weld wherein coalescence is produced by heating to a temperature higher than 800 degrees Fahrenheit by using a nonferrous filler metal with a melting point below the base metals. The filler metal is not distributed in the joint by capillary attraction.

Brinell test   A method for determining the surface hardness of metallic materials.

bronze welding   See *braze welding*.

buildup   The amount of weld face or bead that extends above the surface of joined metals.

burned metal   Term occasionally applied to the metal that has been combined with oxygen so that some of the carbon has changed into carbon dioxide and some of the iron into iron oxide.

burning Violent combination of oxygen with any substance that produces heat. This word is sometimes used in place of the term *flame cutting*.

butt joint An assembly in which the two pieces joined are in the same plane, with the edge of one piece touching the edge of the other.

butt weld The actual weld in a butt joint.

capillary action Property of a liquid to move into small spaces if it has the ability to "wet" these surfaces.

calcium carbide (CaC2) A chemical compound of calcium and carbon usually prepared by fusing lime and coke in an electric furnace. This compound reacts with water to form acetylene gas.

carbon An element that, when combined with iron, forms various kinds of steel. In steel it is the changing carbon content that determines the physical properties of the steel. Carbon is also used in a solid form as an electrode for arc welding, as a mold to hold weld metal, and for brushes in electrical motors.

carbonizing See *carburizing*.

carburizing A carburizing flame is an oxygen/fuel gas flame with a slight excess of fuel gas.

case hardening Adding carbon to the surface of a mild-steel object, and heat treating to produce a hard surface.

castings Metallic forms produced by pouring molten metal into a shaped container or mold.

cathode An electrical term for a negative terminal.

celsius The temperature scale used in the metric system. Zero represents the freezing point of water and 100 is the boiling point (at sea level). To convert to Fahrenheit, multiply by nine, divide by five, and add 32. Celsius was the name of the Swede who invented the centigrade system. Symbol is C.

chamfering See *beveling*.

coalescence Process by which the base metal parts grow together or grow into one body.

coated electrode Metal rod used in arc welding. The rod has a covering of materials that aid in the arc welding process.

complete joint penetration Joint penetration that extends completely through the joint.

concave fillet weld A fillet weld having a concave face.

concave weld face A weld having the center of its face below the weld edges. An indented weld bead.

cone The conical part of a gas flame next to the orifice of the tip.

continuous weld Making the complete weld in one operation.

convex fillet weld A fillet weld having a convex face.

cored solder A solder wire or bar containing flux as a core.

corner flange weld A flange weld with only one member flanged at the location of welding.

corner joint Junction formed by edges of two pieces of metal touching each other at an angle of about 90 degrees.

corrosive flux   A flux with a residue that chemically attacks the base metal. It might be composed of inorganic salts and acid, organic salts and acids, or activated rosins and resins.

coupons   Specimens cut from the weld assembly for testing purposes.

covered electrode   See *coated electrode*.

cracking   The action of opening a valve on a tank of fuel gas or oxygen and then closing the valve immediately.

crater   A depression in the face of a weld, usually at the termination of an arc weld.

creep   The gradual increase of the working pressure (as indicated on the gauge) that occurs because the regulator seat does not close tightly against the inlet nozzle and thus permits the high-pressure gas to leak into the low-pressure chamber. When this condition exists, the regulator should be repaired by qualified personnel before use.

crown   The curve or convex surface of a finished weld bead.

cutting attachment   A device attached to a gas welding torch handle to convert it into an oxygen cutting torch.

cutting flame   Cutting by a rapid oxidation process at a high temperature produced by a gas flame accompanied by a jet action that blows the oxides away from the cut.

cutting tip   That part of an oxygen cutting torch from which the gases issue and burn.

cylinder   A portable metallic container for storing and transmitting compressed gases.

dead-annealed   The result of heating a work-hardened metal to a red color and immediately quenching it in water. This softens the metal and renders it workable again.

deoxidized copper   Copper from which the oxygen has been removed by the addition of a deoxidizer, phosphorus, or silicon. This lowers the electrical conductivity but yields a product more suitable for oxyacetylene welding.

deposited metal   Filler metal that has been added during a welding operation.

depth of fusion   The distance that fusion extends into the base metal or previous layer from the surface melted during welding.

dip brazing   A brazing process in which the heat required is furnished by a molten chemical or metal bath. When a molten chemical bath is used, the bath might act as a flux. When a molten metal bath is used, the bath provides the filler metal.

dip soldering   A soldering process in which the heat required is furnished by a molten metal bath, which provides the solder.

direct polarity   Direct current flowing from anode (base metal) to cathode (electrode). The electrode is negative and the base metal is positive.

distortion   Warping of a metal or metal surface as a result of uneven cooling.

downhand welding   Welding in a flat position.

drag In oxyacetylene cutting, the amount by which the oxygen jet falls behind the perpendicular in passing through the material.

drop-thru An undesirable sagging, or surface irregularity, usually encountered when the welder brazes or welds near the solidus of the base metal. The condition is caused by overheating with rapid diffusion or alloying between the filler metal and the base metal.

ductility The property of metals that enables them to be mechanically deformed without breaking when cold.

edge joint A welded joint connecting the edges of two or more parallel or nearly parallel parts.

electrode A substance that brings electricity up to the point where the arc is to be found.

elongation The total amount of stretching of a specimen produced in a tensile strength test.

erosion Reducing the size of or wearing away of an object because of liquid or gas impact.

expansion Increase in one or more of the dimensions of a body, usually caused by a rise in temperature.

explosion welding A solid-state welding process wherein coalescence is effected by high-velocity movement that is produced by a controlled detonation.

face of weld The exposed surface of a weld.

Fahrenheit A temperature scale used in most English-speaking countries where 32 degrees is the temperature at which water will freeze and 212 degrees is the temperature at which water will boil, at sea level. Symbol is F.

ferrous metals Those metals and alloys of which the principal base or constituent is iron. These metals are magnetic as well.

filler metal Material to be added in making a weld.

fillet To weld metal in the internal vertex, or corner, of the angle formed by two pieces of metal, thus giving the joint additional strength to withstand unusual stresses.

fillet weld Metal fused into a corner formed by two pieces of metal whose welded surfaces are approximately 90 degrees to each other.

flame cutting Cutting performed by an oxygen/fuel gas torch flame that has an oxygen jet.

flanged edge joint A joint in two pieces of metal formed by flanging the edges of the plates at 90 degrees and joining with an edge weld.

flashback The retrogression or burning back of the flame into or beyond the mixing chamber. Sometimes accompanied by a hissing or squealing sound and the characteristic smoky, sharp-pointed flame of small volume. When this occurs, immediately shut off the torch oxygen valve, and then the acetylene valve.

flat position A horizontal weld on the upper side of a horizontal surface.

flowability The ability of a molten filler metal to flow or spread over a metal surface.

flux A chemical compound or mixture in powder, paste, or liquid form.

Its essential function is to combine with or otherwise render harmless those products of the welding, brazing, or soldering operation that would reduce the physical properties of the deposited metal or make the welding, brazing, or soldering operation difficult or impossible.

forehand welding   That method of welding in which the torch and rod are so disposed in the vee that the torch flame points ahead in the direction of welding and the rod precedes the torch.

fuel gases   Gases usually used with oxygen for heating, such as acetylene, natural gas, propane, methoacetylene, propadyne, and other synthetic fuels and hydrocarbons.

fuse plug   A safety device employed on compressed gas cylinders. It consists of a low melting point alloy that melts at a predetermined temperature, thus relieving excessive internal pressure due to heat.

fusion   For the purposes of this book, the melting and flowing together of metals.

gas pocket   A cavity in a weld caused by entrapped fuel gas.

gas welding   A group of welding processes wherein fusion takes place as a direct result of the heat applied with a blowpipe using fuel gas and oxygen. It is always best to be specific when discussing welding fuels; therefore, you will speak of oxyacetylene welding, Mapp/oxygen welding, and so forth.

generator   An apparatus for mechanically controlling the generation of acetylene by the reaction of calcium carbide and water.

gouging   The forming of a bevel or groove by removing material.

groove   The opening provided by a grooved weld.

grooved weld   A welding rod fused into a joint that has the base metal removed to form a V, U, or J trough at the edge of the metals being joined.

hard facing or hard surfacing   The application of a hard, wear-resistant alloy to the surface of a softer metal by an arc or gas welding process.

heat   Molecular energy in motion.

heat conductivity   The speed and efficiency of heat energy movement through a substance.

heat-affected zone   That part of the base metal that has been altered by the heat from the welding, brazing, or cutting operation but might not have actually melted.

heat conductivity   Speed and efficiency with which heat energy moves through a substance.

horizontal position   A weld performed on a horizontal seam.

hose   Flexible medium used to carry gases from regulator to the torch. It is made from rubber and reinforced with fabric.

hydrogen   Considered one of the most active gases. When combined with oxygen, it forms a very clean flame. It does not, however, produce very much heat.

icicles   An undesirable condition where excess weld metal protrudes beyond the root of the weld.

incomplete fusion   Fusion that is less than complete.

inclusion   A gas bubble or nonmetallic particle entrapped in the weld metal as a result of improper torch flame or filler material manipulation.

inert gas   A gas that does not normally combine chemically with the base metal or filler metal.

infrared rays   Heat rays that come from both arc and the welding flame.

inside corner weld   Two metals fused together; one metal is held 90 degrees to the other. The fusion is performed inside the vertex of the angle.

intermittent weld   Joining two pieces and leaving unwelded sections in the joint.

joint   The place where two pieces meet to form a larger structure.

joint design   The joint geometry together with the required dimensions of the welded joint.

joint penetration   The minimum depth of a groove or flange weld extends from its face into a joint, exclusive of reinforcement.

kerf   The space from which metal has been removed by a cutting process.

keyhole   The term applied to the enlarged root opening that is carried along ahead of the puddle in the process of making an arc weld or other type of welded joint.

knee   The lower arm-supporting structure in a resistance-welding machine.

land   The portion of the prepared edge of a part to be joined by a groove weld, which has not been beveled or grooved. Sometimes called *root face*.

lap joint   A welded joint in which two overlapping parts are connected, usually by means of fillet welds.

layer   A certain weld metal thickness made of one or more passes.

lens   A specially treated glass through which a welder can look at an intense flame without being injured by the harmful rays or glare radiating from the flame.

liquidation   The separation of a low melting constituent of an alloy from the remaining constituents, usually apparent in alloys having a wide melting range.

liquidus   The lowest temperature at which a metal or an alloy is completely liquid.

low-temperature brazing   That group of the brazing processes wherein the brazing alloys employed melt in the range of about 1175−1300 degrees Fahrenheit and a shear (lap) joint is used.

malleable castings   Cast forms of metal that have been heat-treated to reduce their brittleness.

manifold   A multiple header for connecting individual gas cylinders or torch supply lines.

manual welding   Welding wherein the entire welding operation is performed and controlled by hand.

Mapp   A stabilized methyl acetylene-propadiene fuel gas often used in place of acetylene.

melting range   The temperature range between solidus and liquidus.

MIG   A term used to describe gas metal arc welding (metal-shielding gas).

mixing chamber   That part of the welding blowpipe where the welding gases are intimately mixed prior to release and combustion.

multilayer welding   In oxyacetylene welding, a technique in which a weld—on thick metal—is made in two or more passes.

neutral flame   A flame resulting from combustion of perfect proportions of oxygen and the welding gas. The most commonly used flame for oxygen/fuel gas welding.

noncorrosive flux   A soldering flux that in itself, and as a residue, does not chemically attack the base metal. It is usually composed of rosin or resin base materials.

nonferrous   Metals containing no substantial amounts of ferrite or iron such as copper, brass, bronze, aluminum, or lead.

nozzle   See *tip*.

orifice   Opening through which gases flow. It is usually the final opening or any opening controlled by a valve.

outside corner weld   Fusing two pieces of metal together, with the fusion taking place on the underpart of the seam.

overhead position   A weld made on the underside of the joint with the face of the weld in a horizontal plane.

overlap   Extension of the weld face metal beyond the toe of the weld.

oxidation   The process in which oxygen combines with elements to form oxides.

oxide   A chemical compound resulting from the combination of oxygen and other elements.

oxidizing flame   A flame produced by an excess of oxygen in the blowpipe mixture, leaving some free oxygen that tends to burn the molten metal.

oxygen   When this gas very actively supports combustion, is is said to be burning; when it slowly combines with a substance, the process is called oxidation, and the result is called rust.

oxygen-acetylene cutting   Cutting metal by use of the oxygen jet, which is added to an oxygen-acetylene preheating flame.

oxygen-acetylene welding   A method of welding that uses for fuel a combination of two gases: oxygen and acetylene.

oxygen cylinder   A specially built container used to store and/or transport oxygen.

oxygen-hydrogen flame   The chemical combining of oxygen with the fuel gas hydrogen.

oxygen hose   See *hose*.

oxygen regulator   An automatic valve used to reduce cylinder pressures to torch pressures and to keep the working pressures constant. They are never to be used as acetylene regulators, and in fact the connections are different.

parent metal   See *base metal*.

pass   Weld metal created by one progression along a weld.

peening   The mechanical working of metal by means of repeated hammer blows.

penetration   The penetration of a weld is the distance from the original surface of the base metal to that point at which fusion ceases.

plug weld   Weld that holds two pieces of metal together. It is made by making a hole in one piece of metal, which is then lapped over the other piece.

porosity   Presence of gas pockets or voids in the metal or weld bead.

postheating   Temperature to which a metal is heated after an operation has been performed on the metal such as welding, cutting, forming, and so forth.

preheating   Temperature to which a metal is heated before an operation such as welding, cutting, or forming can be done on the metal.

psi   Abbreviation for pounds per square inch.

puddle   Portion of a weld that is molten at the place the heat is applied.

quench   To cool hot metal quickly by dunking in a liquid such as water or oil.

reducing flame   An oxygen-fuel gas flame with a slight excess of the fuel gas.

regulator   A mechanical device for accurately controlling the pressure and flow of gases employed in welding, cutting, braze welding, and other processes.

reinforcement weld   Weld metal on the face of the weld in excess of that required for the size of the weld. Its purpose is to add strength.

resistance welding   A process using the resistance of the metals being welded to the flow of electricity as the source of the heat.

reversed polarity   An electrode positive-anode. Referring to dc and causing electrons to flow from the base metal to the electrode.

rod   Metal that welders use as a glue to help join two pieces of metal and that melts at a lower temperature than the metals being joined.

root of weld   That part of a weld farthest from the application of weld heat and/or filler metal side.

safety disc   A mechanical safety device designed for release at a predetermined pressure.

skull   The unmelted residue from a liquated filler metal.

slag inclusions   Nonfused, nonmetallic substances in the weld metal.

slugging   The act of adding a separate piece or pieces of material in a joint before or during welding, resulting in a welded joint that does not comply with the original design, drawing, or specification requirements.

soldering   A means of fastening metals together by adhering another metal to the two pieces of these metals. Only the joining metal is melted during the operation. The joining metal melts below 800 degrees Fahrenheit.

solidus   The highest temperature at which a metal or alloy is completely solid.

spatter   In arc and gas welding, the metal particles expelled during the welding that do not form a part of the weld.

spelter   A term applied to powdered brass used in making a typical brazed joint (lap joint).

spot weld   A weld made between or upon overlapping members wherein fusion might start or occur on the faying surfaces or might have proceeded from the surface of one member. The weld cross section is approximately circular.

straight polarity   An electrode negative-cathode. Connecting dc to cause electrons to flow from the electrode to the base metal.

strain   The reaction of an object to stress.

stress   The load imposed on an object.

stress relieving   Even heating of a structure to a temperature below the critical temperature, followed by a slow, even cooling.

surfacing   The deposition of a filler metal on a metal surface to obtain desired properties or dimensions.

sweat soldering   A soldering method in which two or more parts that have been precoated with solder are reheated and assembled into a joint without the use of additional solder.

tack-weld   A small weld used to temporarily hold together components of an assembly until they can be welded.

tank   See *cylinder*.

T-joint   A joint formed by placing one metal against another at an angle of 90 degrees. The edge of one metal contacts the surface of the other metal.

tensile strength   Maximum pull stress in psi that a specimen is capable of developing.

throat of fillet weld   Distance from weld face to weld root.

TIG   Tungsten inert gas welding.

tinning   In soldering, a coating of the soldering metal given to the metals to be soldered.

tip   Part of the torch at the end where the gas exits and burns, producing the temperature flame. In resistance welding, the electrode ends are sometimes referred to as the tip.

toe of weld   Junction of the face of the weld and the base metal.

torch   The mechanism that the operator holds during gas welding and cutting. At the end of this tool the gases are burned to perform the various gas welding and cutting operations. Often called the blow-pipe.

ultraviolet rays   Energy waves that emanate from the electrodes and the welding flames. The frequency of these rays places them in the ultraviolet ray light spectrum.

undercut   A depression at the toe of the weld, which is below the surface of the base metal.

underfill   A depression on the face of the weld or root surface extending below the surface of the adjacent base metal.

vee groove   See *butt joint*.

vertical position   A type of weld in which the welding is done on a vertical seam and surface.

voltage regulator   An automatic electrical control device for maintaining a constant voltage supply to the welding transformer.

welding   The art of fastening metals together by means of interfusing the metals.

weld metal   Fused portion of base metal or fused portion of both the base metal and the filler metal.

weldment   An assembly whose component parts are joined by the welding process.

weld pool   The small body of molten metal created by the flame of the torch.

welding rod   Wire that is melted into the weld metal.

welding sequence   Order in which the parts of a structure are welded.

work hardening   The increase in strength and hardness produced by working certain metals such as iron, copper, aluminum, and nickel. It is most pronounced in cold welding.

yield strength   The stress measured in psi at which the specimen assumes a specified, limiting permanent set.

Index